WHAT GOES UP

BY JOHN J. NANCE

Splash of Colors
Blind Trust
On Shaky Ground
Final Approach
What Goes Up

WHAT GOES UP

The Global Assault on Our Atmosphere

JOHN J. NANCE

William Morrow and Company, Inc.
NEW YORK

Copyright © 1991 by John J. Nance

All rights reserved. No part of this book may be reproduced or utilized in any form or by any means, electronic or mechanical, including photocopying, recording, or by any information storage or retrieval system, without permission in writing from the Publisher. Inquiries should be addressed to Permissions Department, William Morrow and Company, Inc., 1350 Avenue of the Americas, New York, N.Y. 10019.

It is the policy of William Morrow and Company, Inc., and its imprints and affiliates, recognizing the importance of preserving what has been written, to print the books we publish on acid-free paper, and we exert our best efforts to that end.

Library of Congress Cataloging-in-Publication Data

Nance, John J.
    What goes up / by John J. Nance.
      p. cm.
    ISBN 0-688-08952-6
    1. Global warming. 2. Ozone layer. I. Title.
QC981.8.G56N36 1991
363.73'92—dc20   90-22602
                                                                                                                         CIP

Printed in the United States of America

First Edition

1 2 3 4 5 6 7 8 9 10

BOOK DESIGN BY ROBIN MALKIN

This book is dedicated
with infinite respect and appreciation
to two extraordinary people

RUTH HINSHAW PRIEST CHAPMAN

and

MAJOR GENERAL EDMUND C. LYNCH
United States Air Force (Retired)

Honored members of my family who have redefined
the concept of selfless generosity

AUTHOR'S NOTE

What Goes Up is designed to demystify, in layman's terms, the ozone and global-warming issues, and to give you the real story behind the conflicting, competing headlines and news specials.

But this is also the human story of individual scientists grappling with momentous discoveries and massive risks in a world that doesn't understand their language—and a world that needs to *learn* that language before the miscommunications of the past go too far at some point in the future.

This story is focused on a handful of world-class scientists who have all been as gracious with their time and assistance as they have been squeamish about being placed on my literary pedestal. World-class scientists, you see, never really work alone, and just as an author's single name on a book obscures the fact that many dedicated people were needed to produce it, describing an evolving scientific discovery by following a single scientist can obscure the fact that a network of colleagues helped build the foundation from which the final announcement was made.

A writer, however, has only so much time and space to tell a story and must choose to follow only a few among a field of superlative people, even if those so chosen blink uncomfortably at the light.

CONTENTS

AUTHOR'S NOTE		9
INTRODUCTION		13
PROLOGUE		21
CHAPTER 1.	THE SMOKING GUN	25
CHAPTER 2.	VOICES IN THE WILDERNESS	33
CHAPTER 3.	THE INVISIBLE HOLE IN THE SOUTHERN SKY	66
CHAPTER 4.	IN THE BELLY OF THE BEAST	92
CHAPTER 5.	SHOWDOWN AT THE CFC CORRAL	110
CHAPTER 6.	BURDEN OF PROOF	147
CHAPTER 7.	NOZE-II	163
CHAPTER 8.	THE CORNFIELD MEET	179
CHAPTER 9.	STRANGERS IN A STRANGE LAND	203
CHAPTER 10.	THE VIEW FROM THE GREENHOUSE	222
CHAPTER 11.	A TIME TO CRY WOLF	254
CHAPTER 12.	OF SENDERS AND RECEIVERS	269
	APPENDIX I	283
	APPENDIX II	286
	APPENDIX III	292
	NOTES	299
	INDEX	319

INTRODUCTION

What we have here is a failure to communicate!

Picture a young scientist working in the lab late one night, a dedicated researcher who after months of dogged effort suddenly runs his last series of calculations and finds himself eyeball-to-eyeball with a "monster": If his findings are valid, some long-accepted product or societal practice he has been studying actually constitutes a major threat to mankind—and to his career.

Such a discovery would be galvanizing and momentous of course, but in the place of warm thoughts about Nobel Prizes and professional accolades, cold panic begins to grip the pit of his stomach, steadily encompassing the young Ph.D. with the reality of what probably lies ahead. His head swimming, he sits back heavily, his professional future flashing in horrific images before his eyes. For a brief moment he even considers stuffing the genie back in the bottle—slamming the lid on the Pandora's box before him—rather than face the business community with the conclusion that yet another commodity or activity propelled by the massive inertia of modern society is in reality dangerous to life as we know it on Planet Earth.

Introduction

They will not take it well, and neither will he—his potential "reward" the role of a pariah, the messenger scorned for the temerity of delivering the message.

He knows the precedents, and those who have made such discoveries before him—chemists such as Harold Johnston of Berkeley, Sherwood "Sherry" Rowland, and Mario Molina of U.C. Irvine for instance—soldiers of science who dutifully reported their discoveries about things that gobble the Earth's ozone layer only to find themselves paying a heavy personal and professional price for their trouble. Given their trials and tribulations, our frightened young Ph.D. knows instinctively that he'd better reveal his galvanizing new discovery with the caution of an infantryman lifting his helmet on a stick from the depths of the foxhole, gauging the odds of personal survival by the number of bullet holes he pulls back in.

He knows too that the bullets he'll have to dodge will come first from his fellow scientists, their numbers in direct proportion to the seriousness of his findings. His colleagues will snort, harrumph, niggle, question, retest, deplore, and perhaps decry his conclusions in writing, behind his back, and even in person, and if his "finding" is radical and dangerous enough, they'll fire at him frantically over their shoulders while racing back to their labs like a band of routed Indians in a Western movie—shaken scientists desperate to validate or disprove the "findings" for themselves.

That, of course, is the scientific method of peer review at work, and it does keep science honest. But the process also breeds and licenses human jealousy and professional rivalry, producing at least a few so-called colleagues who will snipe at his work for the pure joy of cutting him down, whether the finding is eventually confirmed or rejected by the overall scientific community.

But he can survive that. As a scientist—a Ph.D. in some scientific field—he's trained to survive the rough-and-tumble world of professional scrutiny. What worries him in the cusp of his midnight discovery is what lies in wait beyond the boundaries of the scientific community, and what happens professionally to those who venture there.

What if there's no one to help him sound a *public* alarm? What if no one outside the scientific community understands the danger and the momentous decisions that must be made by the policymakers representing all of society? What if *he* ends up the one who must go into perpetual motion, abandoning the comfort of quiet scientific research for the glaring lights of public scrutiny, sailing with his findings like the ancient mariner wearing his albatross, trudging from TV stations to Capitol Hill, trying to speak like a scientist with caveats and limitations peppering and ameliorating his statements of alarm, yet trying to trigger responsible action—trying to move the policymakers to value judgments of their own.

Scientists have a hard time speaking the language of the layman when it comes to their work, and the layman has a worse time understanding. If scientists could only be clear, they moan. Is it or isn't it a problem? "When you guys get your act together and agree on something, come back and talk to us."

If Paul Revere had been a scientist, there would have been no armed patriots awake that seminal midnight to meet the British troops. Not even the Minutemen would have understood the cry of a man riding through darkened, rural Massachusetts yelling with appropriate scientific caution:

> I have anecdotal indications that the British may be coming, but I caution that this warning is subject to potential observational error and may be explainable by other phenomena. More research will be necessary before we can state with assurance that the British are, in fact, coming!

And even if a scientifically correct Paul Revere had made himself understood, he would then face the rage of the scientific community for his incautious and unabashed advocacy of a position taken without adequate peer review.

A scientist who takes his findings before the public or Congress runs the risk of being considered professionally biased and suspect. One who dares draw conclusions or value judgments as to how society should respond becomes a radical, and one who crosses the Rubicon to become a visible, vocal advocate for change, may become an outcast—especially if he commits the ultimate sin of publication-by-media (which *is* usually deplorable because such announcements are not tempered by peer review). The purists in science believe their colleagues should state their findings only in cold, detached fashion, drawing no societal conclusions, and making no value judgments or recommendations as to how society should react. But even that is too much for the dwindling number of scientific elitists who believe that scientists should speak only to other scientists and then in professional papers, accepting virtually no responsibility for what use (if any) their less-well-educated brethren on this planet make of their findings.

Science, however, is never pursued in a vacuum of total detachment from the messy realities of the real world. After all, even the ivory towers run on money, and for that money society expects answers in a language it can understand. Pure science and quiet publication is never enough. As life becomes more technological and science more important in every aspect of civilization, we demand ever more loudly that the scientific community come down from their ivory towers every now and then and speak to the common man—and the common politician.

By the same token, the politicians, policymakers, and public have to make a genuine effort to understand the language and the methods of science, if we're to understand their warnings in time.

It took thirteen long, lonely years of sometimes-vicious battles fought by a small cadre of scientists before the world began to limit the ozone-destroying manufacture of CFC's (chlorofluorocarbons). The initial "midnight discovery" was made by two scientists in 1973, yet it wasn't until 1987 that the principal nations of the world reluctantly agreed that the elegant, inert, stable, and highly useful CFC's are indirectly dangerous to mankind.

"But that's a reasonable process," says one respected atmospheric chemist, "when you realize that what we were listening to for all that thirteen-year period were the messy sounds of democracy in action—especially in the United States."

True, yet while that democratic process roared on (and while the CFC industry spent millions vitriolically denying as unproven the scientific worries that their products could be a threat to the ozone layer), millions of tons of additional CFC's were pouring into the sky, threatening additional damage. Time, in other words, is not always on the side of those who would rely on the normal half-life of scientific arguments and the glacial response of political and policy change.

In the end, the argument was resolved as much on the basis of skillful negotiating by a handful of rare and talented scientists and diplomats as by any compelling understanding of the nature of the warning. As the reality of chemical damage—the discovery of the Antarctic "Ozone Hole"—was coming to light, the international community finally acceded to the one truth they couldn't avoid: whether or not ozone depletion *had* occurred in significant amounts, if ever it *did* occur at some point in the future, we would be powerless to stop it for half a century, because the CFC's put in the air today will take fifty to sixty years to even reach the upper atmosphere. There had been far more compelling facts that should have spawned a worldwide ban of CFC production immediately in 1975, but such findings were largely swept aside because the scientific community was not "certain."

Even today the international response is wholly incomplete, and CFC's continue to waft skyward. The messy sounds of democracy in action may be joyous in theoretical debates, but when we're fouling our own nest, we may not have the luxury of waiting out the process to see what happens.

It has taken forever, it seems, for governments and politicians and the chemical industry to realize that as simple as it sounds, what goes up must go somewhere, and that somewhere in this case is the atmosphere that keeps us alive. And logic alone, it would seem, would tell us that blind, wholesale alteration of that gaseous soup can't really be a good idea, especially when we don't fully understand the consequences now and for the future.

The Chlorofluorocarbon/Ozone battles involved a single family of man-made gases, which can be replaced. The far more serious global threat that has taken its place in media attention—the so-called Global Warming caused by The Greenhouse Effect—is a monster of a different temperament. The gases involved are many, and stem from the very essence of modern society: the production of energy for human use. There are no obvious and simple methods of turning off or turning down the worldwide overproduction of carbon dioxide, methane, and nitrous oxide (and other trace gases), and there is major uncertainty over the consequences of ignoring the problem, or being too slow to act. Do we have another thirteen years for *this* debate? No one knows as yet, and the answer may be a shocker.

In the unusually hot summer of 1988 in the United States a single scientist touched off a worldwide firestorm of publicity and scientific recrimination by expressing a personal conviction: that we were already beginning to pay the wages of atmospheric sin—increasing greenhouse gas emissions over the past one hundred years—through the sizzling temperatures and Midwestern drought conditions. A massive scientific catfight erupted as a result, a battle that has obscured and twisted the basic message: Whatever the eventual effect on the climates of the world and the levels of the oceans, no credible member of the atmospheric science community disagrees with the evidence that we *have* altered our atmosphere rather dramatically (in terms of the mixture of gases), and we face uncertain consequences as a result.

In many respects the Global Warming/Greenhouse Effect debate is at the same crossroads occupied by the chlorofluorocarbon debate in 1974—scientific uncertainty fueling societal and political reluctance to act. In the process, a monstrous hurricane of sound and fury has obscured the real messages coming from the scientific community, and no one is really communicating. The public and the media have been watching a play-by-play account of a controversy that really isn't significant, focusing on a side-show battle over whether or not we are already experiencing measurable global warming. Was the summer of 1988 the opening round? Was the decade of the eighties the hottest one in a thousand years? The fact that those questions don't really matter has been temporarily lost in the noise of stories and articles reacting to the slightest cooling trend with headlines questioning whether the Greenhouse/Global Warming alert is a false alarm.

It isn't, yet we've missed the point. The important question does not depend on whether or not the climate has begun to change. The important question is simply this: Just how far *should* we go in permitting continued uncontrolled experimentation with the life-giving atmosphere of our home planet—the only habitable planet we know of in this or any other galaxy.

Compared with the size of the Earth that nurtures us all, an individual human seems insignificant. How could such tiny creatures as we affect in any material way such a huge planetary body?

Yet we *have* affected it. Untold billions of free chlorine atoms now float in the stratosphere for the first time in Earth's four-billion-year history, and 25 percent more carbon dioxide now occupies the gossamer-thin atmospheric envelope than existed a mere century ago. Billions of human beings—our numbers growing by the minute—spew chemicals and gases indiscriminately into the sky, and only within the past two decades has anyone seriously questioned the long-term consequences.

"I can't go home and dump my garbage in my neighbor's backyard," says atmospheric chemist Dr. Susan Solomon, "the police would arrest me in five minutes. But I could take a tank of chlorofluorocarbons, put it in my backyard, turn it on and let it go into the atmosphere all day long, and no one can stop me! Somehow that's very wrong."

The point is that mankind faces for the first time the need to make some very weighty policy decisions for all of society. And yes, those decisions depend on what the scientific community believes to be true.

But here's the problem: If we can't understand each other, if we can't decipher the warnings, we simply argue before a Tower of Babble, flailing our arms in animated debate delivered in a foreign language beyond the comprehension of our fellow debaters, accomplishing nothing—while the hot sun gets hotter above us.

There will be future midnight discoveries. It's up to all of us within and outside the scientific community to learn to speak with a common tongue when we face such threats.

PROLOGUE

A faint glow of reddish-orange brushed the eastern horizon where moments before only darkness could be f nd. With each passing second as the planet turned inexorably to the east, more of the wildly scattered rays of light from the distant sun pushed through the thin lens of Earth's atmosphere, forming a slight line of ruddy color, heralding the impending sunrise. It would be minutes before the mighty solar engine of electromagnetic energy would be fully exposed, and before the incoming radiation from ninety-three million miles away could have a straight shot at a tiny puff of oxygen that floated in the frigid Antarctic darkness. It was an impossibly small formation of identical molecules in a sea of stratospheric gases at sixty thousand feet above the southern ice cap, adrift in a soup of floating molecules sitting like a carefully constructed amalgam of dry tinder and twigs among logs and planks, waiting for a spark to turn it into a raging bonfire of chemical reactions—a spark now approaching over the eastern horizon in the form of sunlight in various wavelengths. The submicroscopic community of oxygen atoms had long since bonded together in threes, becoming part of a vast atmospheric reservoir of similar molecules called ozone—a gas at once poisonous to humans yet indispensable for human life. And, for the first time in tens of millions of Earth years, an endangered species.

These particular oxygen atoms had traveled a great distance together, beginning two years before at the southern tip of New Zealand as they percolated into the sunlight through the cells of an aspen leaf. They had come out in pairs, two oxygen atoms clinging to each other to form a molecule known as O_2—the stuff of carbon-based life itself. Each of the tiny pairs had once been married to a carbon atom, but the sunlight and the strange inner workings of the aspen leaf had stolen the carbon and pushed them into the open to float up and away in a sea of fellow oxygen molecules and nitrogen gas, eventually drifting toward the southern pole.

And it had been a year since the ultraviolet sunlight had come over the eastern horizon on a similar morning, energizing an age-old process: breaking the bonds holding many of the tiny pairs of oxygen atoms together. In untold numbers they had begun to split into single atoms, which then floated in aimless solitude until bumping into a surviving pair whose bonds had not yet been torn away. When such collisions occurred, the pair would attach in a microsecond to the single atom, and the three of them would spin off in another direction, now with the different characteristics that come with forming O_3—the pale blue gas known as ozone. Hour after hour, day after day, the cycle continued, the ozone molecules absorbing a special wavelength of ultraviolet light, UV-B, which would break the bond holding the three together and cause them to split off a single oxygen atom which then promptly attached itself to an O_2 molecule, once again forming an ozone molecule.

There was a beautiful symmetry to the delicate molecular dance, a self-regulating system that kept the gossamer veil of ozone in proper supply; there were always enough free oxygen atoms to reform the ozone that continually sacrificed itself to the cause of absorbing ultraviolet energy.

Now, however, something new and threatening had joined the atmospheric community, something lurking in man-made abundance as the sun's rays cascaded over the eastern horizon, ending the six-month Antarctic night. Chlorine atoms, untold trillions of them, unnatural, foreign interlopers in the stratospheric equation, rapidly became energized by the rising level of sunlight, and began their own catalytic dance. In accelerating numbers, each chlorine atom would rip away a single oxygen atom from an ozone molecule and transform itself into a molecule of chlorine monoxide (one chlorine and one oxygen atom bonded together). What had been an ozone molecule now became an O_2 molecule again.

But the process was just beginning. The chlorine monoxide molecules now began a complex and unique metamorphosis, pairing off to form two ClO molecules together—a "dimer"—mutating eventually to end up as single Cl atoms on one hand and stable O_2 molecules on the other. The net result in each cycle left each chlorine atom just as it had been at the start, but for each ozone molecule caught up in this strange polar catalytic cycle, there would remain only stable O_2 molecules which would split no more, and form no more ozone.[1]

As the sun rose higher, the unnatural collection of chlorine scavengers began moving faster, each one destroying ozone molecule after ozone molecule, endlessly repeating the same cycle as more of the ultraviolet radiation filtered deeper and deeper through the stratosphere toward the unprotected surface of the Antarctic ocean twelve miles below—fewer numbers of ozone molecules left to intercept the damaging frequencies. Day after day the process proceeded until by mid-

October—the Antarctic spring—nearly 80 percent of the ozone molecules had been destroyed in a slice of the stratosphere between sixty thousand and one hundred thousand feet above the south polar region.

Nothing like it had happened before in such magnitude, and in the silence of space overhead, a man-made satellite took careful note of the disappearance and the deepening "hole," digitalizing what it saw into long strings of radioed binary messages sent back to the United States to a high-speed computer—data strings that painted a picture of ozone concentrations so ridiculously low that the computer did precisely what it had been programmed to do: It rejected the data as obviously incorrect.

CHAPTER

1

The Smoking Gun

Cambridge, England, October 1984

Joe Farman leaned back in his chair and shook his head slightly in resigned wonder, the gesture unseen as he sat among the wooden shelves and slightly overstuffed file cabinets of his rather plain office, still stunned by what he and several colleagues had decided in an Advisory Group meeting several hours earlier.

Beyond his window, the city lights twinkled in the grip of a cold October evening, the comparatively mild temperatures mocking the conditions eight thousand miles to the south in Antarctica at the British scientific outpost—a place where even a warm day could give you frostbite.

His thoughts wandered there now, reexamining the data yet again. It was "spring" in Antarctica, though you'd never know it from the subfreezing temperatures. In September and October of each year, as the anemic sunlight increased in intensity (giving birth to little more than slightly warmer ice in the permanently frozen wasteland), ships began arriving with fresh teams of scientists to continue a dizzying array of studies at several outposts on "the ice"—and Joe's colleagues of the British Antarctic Survey were always among them. For twenty-seven years it had been so, but not without a constant struggle for funds.

He didn't miss being there with them this year. That was for younger men. Not that he was getting old, of course. Tired, maybe, and perhaps a bit bored, though that would certainly change now, thanks to the latest readings. The printouts had landed like a private bombshell hours before, confirming a secret he had been sitting on for three long years.

Farman looked again at the papers spread on the desk before him—papers containing readings of ozone levels in the stratosphere now 40 percent below normal, an ominous fact that had triggered his decision to call the Advisory Group together. The trend seemed far too clear to be ignored any longer—and one of the group, atmospheric physicist Bob Murgatroyd, had agreed immediately. They would have to release the findings—publish a paper containing the amazing data—and try to explain it at the same time. That much, at least, *was* clear. The professional consequences that might follow were not.

Joe replaced the papers and sat back yet again in deep thought, his slightly weathered face catching the light at an angle fit for black-and-white portraiture. Of medium height and build, Joe was usually in perpetual motion, a bundle of nervous mannerisms and rapid speech tinged with good humor and courtesy all contained in a slightly rumpled suit and set off by a perpetual cigarette, but he was uncharacteristically still now as his mind ranged over the problem.

Three years before he had all but dismissed readings flowing north from his Halley Bay team, which showed lower ozone levels than the year before. In 1982 the readings had dipped even more obviously, yet it was *still* too soon, or so he felt. The next year brought a further decline, prompting Joe to add another monitoring station to watch the ozone concentrations at a spot nearly a thousand miles northwest of Halley Bay. They had been all set for 1984—set to have additional control over the data, to help ensure that they wouldn't end up in the position of scaring the world with inaccurate measurements. The possibility that the small team of handpicked scientists could become a world-class target of professional guffaws and derision for publishing "bad science" haunted him, like a recurring nightmare.

Joe Farman knew he was walking a razor-thin line, and he hadn't struggled for nearly three decades to keep the Survey going only to let it be blown into obscurity by releasing a premature report. Yet it went against the grain as a scientist to sit on the findings. He had even had to block the excited requests of a young Ph.D. candidate a year earlier when the fellow had grown wide-eyed over the downward trend and wanted to publish the data in his thesis. It was too soon, Joe had explained. After all, even if the figures weren't instrument errors, the lower ozone readings might be the result of some natural process, some unknown variation of nature.

Deep down he had known that couldn't be the case. Deep down he had felt for a long time that man-made chlorofluorocarbons were somehow at the root of the problem. But that was a leap of speculation, not a scientific finding. They would need a warehouse of confirmed findings before he spoke of such suspicions in public.

Joe thought back over the decades he had been working in, and around, Antarctica. He had been an eager young scientist during the IGY—the International Geophysical Year of 1957, which was a seminal gathering of geophysicists and scientists from various disciplines which had launched many a career. Measuring ozone and other things in the frozen wasteland of the southernmost continent had become a completely routine exercise even before the so-called ozone wars broke out in the early and mid-seventies—battles over the use of spray can propellants and other uses of man-made gases called CFC's, or chlorofluorocarbons. All the political and environmental hubbub of that tumultuous period had focused worldwide scrutiny on the ozone layer and its role in protecting life on the surface from damaging forms of ultraviolet light (UV-B radiation). Tens of millions of tons of CFC's had been released into the air, and the molecules—each containing chlorine atoms—were wafting their way skyward like an invading air force of tiny bombs, each of them filled with the potential to destroy large numbers of ozone molecules. So vicious and personal were the battles between scientists and industry during the seventies that by the time they were effectively "over," the whole issue of CFC's affecting the ozone layer had become politically supercharged—a built-in threat to anyone who might have the misfortune of being the first to actually "find" a decrease in global ozone levels. Whoever *was* first with a downward trend, Joe knew, would find himself under excruciating professional scrutiny as a globe full of fellow scientists rallied to the task of disproving both finding and finder.

Joe Farman had watched and waited for someone else to climb into that arena as his readings from Halley Bay continued to suggest "something" was happening to the ozone overhead. After all, he had a research outfit to protect.

The British Antarctic Survey was all but invisible even within the British scientific community, and the Halley Bay outpost was equally unknown. To Joe Farman, however, the Survey had become professional life itself, and he had protected it like a mother lion, arguing endlessly over the years about the inestimable value of providing the rest of science with a trustworthy baseline of atmospheric data stretching back decades. Since 1957 he had labored to keep them going even as members of the British government tried to cut his funding. They had run the whole thing on four hundred thousand pounds per year (less than a million in pre-eighties U.S.

funds), and Farman himself had subsisted on the equivalent of less than twenty thousand dollars annually. It was hardly a glamour outpost of science. Yet to Joe Farman, the mission was worth it.

The academics and the government political leaders who had tried to quash the Survey in past years were still out there, and a major stumble or embarrassment could give them the ammunition they might need to administer a coup de grace.

Six months before Joe had participated in briefing a famous scientist, a Cambridge don who was also the director of the National Environmental Research Council—and Joe's employer—on the importance of understanding how the atmosphere of the world is changing, and how, to gauge that change, careful measurements over a long period of time were essential, even if somewhat boring. Joe excitedly told him about the trends they were tracking in ozone, and about the ranking suspect, chlorofluorocarbons, thinking the effect would be positive.

"Why on earth are you measuring *that*?" he demanded.

Joe explained the immense volume of the CFC's being produced, and the danger to the ozone layer worldwide if they were in fact harmful—an effect they could only prove if they could monitor a decrease in ozone. "That's what we're doing," he volunteered, watching the man for clues to his reaction.

The director—an aristocratic gentleman with considerable power to destroy or create funding for scientific projects by little more than a phone call—suddenly straightened up and looked at the group of eager-to-please scientists, including Joe Farman, who watched with apprehension as a light came into the director's eyes, the chance to toss off a particularly clever remark becoming irresistible.

"Oh, so you're making these observations for posterity, eh?"

With the towering self-confidence befitting a Cambridge don of high standing and secure future, he looked at the comparatively lowly defender of the Antarctic Survey, and tossed a small hand grenade into Joe's psyche.

"And what," he asked in impeccable English tinged with a smug sarcasm, "has posterity done for you?"

It was a casual comment, of course—a vain attempt at humor without any understanding of its impact on the small group of scientists before him. Joe Farman knew that intellectually. But the effect was akin to that of a king who absently flips a gesture of dismissal at a lowly subject—whom his faithful courtiers promptly execute.

Such statements have stopped scientific research in its tracks countless times in history, and in this case it had cost Joe's group several months of nagging uncertainty and delay. After all, he was about to put the reputation of his carefully nurtured Survey on the firing line, and if they were wrong, if they drew too much fire, those who had little understanding and less support—the people he had quietly battled against for years—might finally pull the rug from under them all. There was, in other words, far more than posterity at stake if the trends were valid.

These latest figures—from both Halley Bay and the second observation post—had broken the back of his hesitation. First or not, they had to publish.

Joe worried about the inevitable charges that his instruments were merely out of calibration. The spectrophotometers his people used to look at the patterns of light penetrating through the various molecules over Halley Bay were old units that required periodic recalibration.[1] They could drift off. They could be wrong, as the rest of the atmospheric scientific community would rush to point out.

He chuckled to himself, thinking of the standard response of his team and most everyone else who measured ozone levels in the stratosphere. Whenever there seemed to be a dip in the data—a downward trend in the amount of ozone overhead a particular spectrophotometer—no one would believe it! The instrument's calibration or the math involved in the calculations was immediately suspect. "My God, it looks like a trend! Okay, now where did we go wrong? Obviously the data are screwed up!"

That had been his first thought several years back when the first visible decline in total ozone showed up in the data from Halley Bay. Instrument error. Calibration problems. Observational mistake. "*Something's* bloody well screwed up, because this *can't* be a trend."

But now he was sure of his data—which gave rise to another chilling question: Why hadn't anyone else in Antarctica seen the decline in ozone readings? The Japanese were on the ice at a different location, weren't they? So were the Americans, and there were satellites orbiting overhead peering into the same mass of cold air. Why the hell hadn't *they* recorded anything like this?

Yet here he was in 1984 with a fait accompli: Something was voraciously destroying the ozone above his people, and it would be up to Joe Farman, Ph.D., defender of his beloved Antarctic Survey, to report it. What lay ahead would undoubtedly be about as exhilarating as a morning run through a minefield.

Joe Farman looked at his watch and pulled himself out of his chair, thinking of home. His tall, thin frame was draped with casual clothes and a sweater that made him look slightly disheveled, which was probably not incongruous. Tonight he felt somewhat disheveled, especially as he turned his considerable intellect to the task of explaining the phenomenon of the missing ozone.

The suspect was chlorine, of course. But how could they prove it? All he had were uncorroborated readings in the range of 150 Dobson Units (where they should be seeing close to 300 for a normal ozone layer).

Joe knew of the work of Drs. Sherry Rowland and Mario Molina in the United States, the chemists who in 1974 had first sounded the alarm about putting too much free chlorine in the stratosphere, warning that CFC's could break down at higher altitudes and release their chlorine atoms to combine with ozone molecules, reducing them to stable oxygen (O_2) pairs. But those theories concerned the *entire* ozone layer covering the entire Earth, and the worldwide measurements of ozone hadn't shown it was happening in any quantity. More important, no one had theorized that the levels of ozone could drop *exponentially*—in ever-increasing percentages.[2] Any damage to the ozone layer should have been slow and steady in predictable increments—a *linear* process. What was happening in Antarctica was clearly *non*linear, and that seemed to make no sense at all.

Rowland and Molina's work had helped slow CFC production, but there was still no hard proof that the increasing chlorine population above ten miles in altitude was going to have a significant effect. There was, in other words, no smoking gun.

Had the little British Antarctic Survey—often ignored, intimidated, and threatened by a disinterested Parliament and an academic community that found its function rather boring and without purpose—done exactly that: found the smoking weapon?

Joe Farman and most every other scientist connected with atmospheric studies could remember only too well the thundering response of an indignant chemical industry on several continents to the 1974 Rowland-Molina paper that indicted CFC's as a threat to ozone, and a potential instrument of increased skin cancer and crop damage. "Preposterous!" industry had raged, trotting out an impressive array of company chemists and public relations people flanked by political lobbyists all trying to make the point that there was no proof of ozone depletion in the sky, and without proof there was no reason to damage a money-making industrial product—the eight-billion-dollar-per-year CFC industry!

"Where the hell is the *evidence*?" they had demanded.

In the stratosphere over Halley Bay, apparently, Joe Farman realized.

CHAPTER
2
Voices in the Wilderness

University of California at Irvine, December 1984

As Joe Farman began writing up the frightening figures of ozone loss from Halley Bay, six thousand miles to the west Dr. Sherwood Rowland was facing the reality that the CFC battle was all but lost. No one was listening, and a million tons per year of chlorofluorocarbons continued to churn into the atmosphere from an unconcerned world.

"The ozone layer problem? *What* problem? Didn't they solve that ten years ago?"

Rowland was not about to give up, of course. It wasn't in his nature. For that matter, neither was bitterness. Whatever frustration he felt—and he *was* frustrated—was contained beneath the even demeanor of his friendly smile and his eternal willingness to patiently explain even the simplest of things to

[33]

the simplest of reporters, though fewer of them came around these days.

Rowland was a bear of a man, six feet five inches tall with huge feet, thinning hair, a broad, pleasant face—and the totally incongruous nickname "Sherry," a permanent touch of humor that often caught the newly introduced off guard.

"Uh, *Sherry* Rowland?"

"Better than Sherwood."

Not that Sherry wasn't known to the world of chemistry—and for the past eleven years, more than well known to those who deal with the atmosphere in various disciplines (atmospheric chemists, dynamicists, aeronomists, physicists, etc.). After all, it had been Sherry Rowland and his (at that time) postdoctoral research associate Dr. Mario Molina who had innocently stepped up to the bar of public opinion and political reality in 1974 and laid before the world an unbelievable theory: One of the most brilliantly conceived of chemical products—the exquisitely stable molecules known as halocarbons (chlorofluorocarbons), which could energize refrigerators and car air conditioners, blow plastic foam into useful shapes, clean electronic circuit boards, and safely propel the ingredients of underarm deodorant and hair spray cans—hinged on a deal with a chemical devil, the price for which could be the soul of the ozone layer several generations into the future.

"The *what* layer?"

Not too many people outside of atmospheric science, aeronautics, and astronautics had even heard of the ozone layer. A seemingly ethereal collection of bluish gas composed of three oxygen molecules bonded together in a very tenuous, easy-to-shatter ménage à trois, the ozone layer floats somewhere between eight and twenty-five miles above the planet, so those who were interested enough to ask were told. But in 1971, the world was presented with the sudden and well-publicized worries of several scientists that the exhaust gases from a worldwide fleet of high-flying supersonic transport aircraft (SST) could do substantial damage to the ozone layer, which would increase skin cancers among humans on earth. It was a new concept, and a new global threat to the environment. Suddenly TV networks were doing specials on the subject, and new, colorful descriptions were coined for the scientifically uninitiated. As one narrator explained it, "If you compressed all the ozone around our planet to normal sea level atmospheric pressure, the thickness of that layer would be only three tenths of a centimeter—around an eighth of an inch thick—an impossibly thin membrane of poisonous gas that, ironically, protects all life down here below." The public was told in various articles that the strange little molecules absorbed a type of ultraviolet sunlight (UV-B) that was harmful to human skin and the DNA molecules common to all living things on Earth.

It was the fight over the alleged linkage between skin cancer and airline equipment that first brought phrases such as "ozone layer" and "nitrogen oxides" into the language and public consciousness (nitrogen oxides were the worried-about molecular compounds left in the wake of high-flying jet engines that were thought to cause a catalytic reaction with ozone and lower the protective level of the gossamer-thin ozone layer).[1] But when the Boeing SST program was killed in 1971 (more for political and economic reasons than for any concern over ozone), the public's attention—and that of their senators and congressmen—largely drifted to other things. Richard Nixon had just accepted his second coronation as president, and a "third-rate burglary" at the Watergate in Washington was about to rivet the attention of the nation for several years to come. After all, there were no SST's being built, so there could be no threat to the ozone layer, right?

The atmospheric scientific community was not so easily convinced. Respected scientists such as Dr. Harold Johnston of Berkeley—a Southern gentleman of considerable charm and a steel-willed determination not to be cowed by political pressure—had been taking flak for several years for his statements about SST's and their potential for damaging the ozone layer. (He would later take great pride in finding he had been placed on President Nixon's "enemies" list for his opposition to the SST.) Though he handled it reasonably well, there was damage. A scientist's lifeblood is his reputation, and Johnston's had been sullied. He was an exceedingly careful senior chemist of stern ethical fiber, but he had been portrayed as a hotheaded radical for simply following his conscience and reporting his scientific conclusions. There was a bitterness in Johnston over it all, an edge that those who studied under his direction saw from time to time—young scientists such as Dr. Susan Solomon, who would later face similarly rarefied political/economic debates with the advantage of having Harold Johnston's experience to draw on. By 1984, Harold Johnston's battles had been all but forgotten outside the atmospheric chemistry community. But in 1973—when Sherry Rowland found himself pulled into the arena of the ozone debates for the first time—the wounds were still very fresh.

University of California at Irvine, December 1973

The sight of the quiet and contained form of Mario Molina hunched over his desk in the midst of a large office housing several other postdoctoral assistants was not unusual. He had held a similar pose many times over the previous weeks and months, working away with pencil, paper, and electronic calculator to chase down the answers to an interesting pure-science problem he had picked from among those Sherry Rowland had offered him. But this was not making sense. In fact, in some ways, it was simply too bizarre to be true. There *had* to be a rather monstrous error—but where?

Mario attacked the figures from a different angle, unaware that the speed of his numeric entries had increased just slightly, the clickety-clacking of the calculator keys barely audible, and all but unnoticed, in the clutter of the room.

Sherry Rowland had been quite pleased when Molina, a newly minted thirty-year-old Ph.D., agreed to join his group of postdocs at the relatively young Irvine campus. Irvine was a typical scientific teaching outpost, headquartered in a large, rectangular, six-story box of a building, and Molina's new position was blessed with comfortable anonymity (if modest pay).

Mario Molina came highly recommended from the Berkeley chemistry department. The son of the Mexican ambassador to the Philippines, he was an exquisitely educated young scholar of quiet demeanor, his closely cropped beard and short stature complemented by a quiet voice, intense eyes, dark hair, and kindly, almost shy appearance that seemed incongruous when he got to his feet to ably defend some point. The physical contrast between Mario and Sherry Rowland was almost stark. Rowland's imposing height and physique, his resonant voice and great stride were a counterpoint to the smaller measurements, quieter voice, and slight limp (from a youthful soccer injury) that characterized the new addition to Rowland's team. Yet Molina and Rowland were intellectual soul mates, both of intense intelligence and world-class scientific standards which held the promise of fruitful collaboration.[2]

Mario was not looking for notoriety any more than Sherry Rowland. But if these figures . . .

That was it. He had hit the wall. Mario Molina stuffed some of his calculations and papers in a folder and charged out of the office in search of Sherry Rowland. There was an error here someplace and he needed help in finding it.

The question they had decided to answer was quite simple and straightforward, uncluttered by political considerations or societal impact—or so Rowland had thought. "Where," he had asked, "do chlorofluorocarbons go when they escape at ground level?" When Freon (Du Pont's brand name for their premier CFC product) escapes during a recharge of a car air conditioner, or the halocarbon propellants are released from spray cans, where, and how high, do those molecules go before starting back down in some form? Chlorofluorocarbons are so stable they won't react with any product they propel or affect the skin of any human being they touch, but they have to go *somewhere*.

The CFC molecules were tiny packages of carbon, fluorine, and chlorine atoms bonded together which simply begin mixing with the rest of the atmospheric soup when released to the sky. Over time, the tiny molecules are swept up in the large-scale circulation of winds that carry them upward—a process known as turbulent diffusion.[3]

But precisely *when* do they come *down*?

Mario had chased equations around with his pencil and calculator until he was, quite frankly, a bit bored by the whole thing. Every possibility for the halocarbons to react in the Troposphere—the layer of atmosphere from the Earth's surface up to the thirty-five to forty-five-thousand-foot level, and the one in which almost all weather phenomena occur—ran into a dead end. Halocarbons were tough and resilient, and chemically snobbish. They wouldn't have anything to do with other molecules in the Troposphere, as far as he could tell, and there wasn't enough ultraviolet light from the sun coming through at those altitudes to break the molecules apart.

In effect, it was like a sky full of tiny plastic chips: The damn things were indestructible—at least in the Troposphere. They didn't react, they didn't "rain-out" (get swept to the ground by raindrops) because they wouldn't dissolve in water, they just stayed there, moving slowly upward in the circulatory patterns of global winds and turbulence.

Toward the Stratosphere.

Both of them had thought of stopping at that point and writing a paper on their findings. After all, they had answered their initial question quite thoroughly. Chlorofluorocarbons simply didn't break up in the Troposphere. In the language that normally described natural processes that remove polluting gases and particles from the atmosphere, they had been looking for a Tropospheric "sink," something that would in effect pull the CFC's from the atmosphere to end up harmlessly on the Earth's surface. But there was no sink for CFC's, other than the Stratosphere. They were now sure of that.[4]

"Well, we might as well be complete about it," they decided. There were chlorine atoms involved, and since they knew that chlorine was very reactive and would find several compounds in the stratosphere to react with, the question was *which* compounds at stratospheric temperatures and concentrations would attract the reactive attentions of the chlorine atoms released from CFC compounds. Among other molecules, ozone occupied those altitudes.

Sherry Rowland was not an atmospheric chemist, nor was Mario Molina. Neither was trained in dynamic theories of air transport, or meteorology. They were chemists. This airborne world was new to them and interesting because of that newness. Mario had begun his detective work by reading up quietly and thoroughly on things atmospheric, and was sufficiently sophisticated to know the nature of the Tropopause, the boundary layer that separates (and to some degree insulates) the Troposphere from the Stratosphere above. He was well aware, too, that in absorbing damaging ultraviolet light, ozone also absorbed heat, which was one of the principal reasons the temperature of the Stratosphere does some strange things: As altitude increases, the Troposphere gets colder, but the Stratosphere gets warmer.

If halocarbons didn't break up, fall out, or wash out of the Troposphere, they would continue being dispersed (by turbulent diffusion) into the Stratosphere, and eventually—when they were high enough and there was too little ozone above to filter out the extremely energetic wavelengths of UV radiation from the sun that can break the CFC molecular bonds—those bonds *would* be broken by photodissociation, and the once-indestructible man-made molecules would be reduced to various shards and pieces of molecular and atomic particles, including chlorine.[5]

In fact, chlorine, Mario determined, would end up within an interesting little molecule called chlorine monoxide (ClO).[6]

ClO is a free radical, which means it's promiscuous—it will mate indiscriminately with almost anything—especially another unstable molecule. (A free radical has an odd number of electrons, an essentially unstable arrangement that causes it to search endlessly for a way to react with other atoms or molecules.)[7]

When ClO collides with a free oxygen atom, it "gives" *its* oxygen atom to the stranger, forming a very stable marriage (O_2) with an even number of electrons. The newly formed O_2 goes floating happily away, leaving in its wake a single chlorine atom. And now the chain reaction begins.[8]

One chlorine atom, Mario discovered, would react with ozone, destroying successive molecules of ozone in a simple series of steps that always left the same chlorine atom in its original condition.

Chlorine is rapidly attracted to unstable ozone molecules (if there are any in the gaseous neighborhood), and on meeting one, will effectively steal one of the oxygen atoms from the ménage à trois. What was then an encounter between a single chlorine atom and the triple-oxygen formation known as ozone (O_3) now transforms to *another* stable O_2 molecule, and yet another promiscuous chlorine monoxide (ClO) molecule on the prowl for another free oxygen atom. When the ClO then hits an O, the product is a stable O_2 and a free chlorine atom once more. Like a fast-moving bachelor with a firm intention to avoid marriage, the chlorine transitions through two molecular relationships in each cycle, destroying ozone, leaving copious colonies of stable O_2 behind, always returning to its free state as a single chlorine atom on the prowl. (ClO + O → Cl + O_2, then Cl + O_3 → ClO + O_2, and finally, the ClO from that reaction: ClO + O → Cl + O_2).

The loser is ozone. The peripatetic chlorine atom never meets an ozone molecule it doesn't destroy. It becomes the agent of destruction of ozone—a catalyst—but true to the definition of a catalyst, it promotes change without being changed.

Mario Molina worked through this reaction, though he had found nothing in the scientific literature about it.[9]

Okay, he figured, so we have a catalytic reaction in the stratosphere and ozone is the loser. Ozone is *always* being broken down and reformed anyway (O_3 bonds are broken by ultraviolet light leaving O_2 and O, which promptly hit similar partners to re-form O_3). So a few wild chlorine atoms take out a few thousand ozone molecule before they finally attach to some more complex molecule that takes *them* out of the picture (a sink at last for chlorine).

So what? The amounts were trivial, because the amount of chlorine was trivial. There were measurements showing only a tiny amount of CFC's actually in the stratosphere, though, of course, there was more of the stuff on the way up.

How much of the CFC's had been produced, he wondered? It seemed like an important question since, eventually, everything produced would reach the Stratosphere. Molina spent several days chasing down figures on the volume of chlorofluorocarbon production worldwide, and it was then that his figures began doing strange things. He had expected those figures to be trivial as well.

They weren't.[10]

"Sherry, we have to talk!" Mario caught up with a rumpled-looking Sherry Rowland in the hallway. His senior was getting ready to leave for a five-month sabbatical in Vienna—a Guggenheim fellowship he and his wife had been looking forward to—and the process of getting ready to go was obviously a strain. On top of it all, only about two weeks remained before Christmas.

But Sherry Rowland always had time for those who wanted to talk to him. The two scientists returned to the office and Mario explained the problem. Sherry knew instinctively there had to be an error, just as Mario did. After all, no one else had sounded any alarms about chlorofluorocarbons in the stratosphere, but the industry data showed that vastly more CFC's were being produced than any other trace gas that was known to be reactive with ozone. Obviously, the erroneous figures on Mario's pad indicated that the catalytic reaction of chlorine multiplied times the volume of chlorine that would eventually be injected into the Stratosphere by CFC's would have global consequences.

They wore down their pencils as they worked the figures over, punching the numbers into their calculators and retracing every step, looking for the big error that continued to elude them.

The next day they tried again, and by the end of the session—with no verbalization of the consequences of what they seemed to be seeing—they decided the error either didn't exist, or was too buried to find.

Sherry Rowland knew of Dr. Harold Johnston's deep involvement in the questions of ozone and atmospheric reactions, and he phoned him, outlining that they had found a chain. Rowland was somewhat startled to hear that the catalytic chain reaction was already known. What had *not* been found, said Johnston, was a sufficiently large source of chlorine in the Stratosphere to cause any real worry. There had been some concern that the space shuttle might inject chlorine, but those amounts seemed insignificant as well.[11]

"Well, we've found a chlorine chain *and* a chlorine source ... we're trying to find out how much of this is known within the atmospheric science community," Rowland told him, and on December 27 they met in Johnston's Berkeley office.[12]

Harold Johnston went over the Rowland-Molina figures in great detail before informing the two men that they had indeed just found the first sizable source for free chlorine in the Stratosphere, and that he could find no errors in their work. Johnston knew what was ahead for the two. He had been through the mill already with the SST–ozone fight, and was still feeling the effects. Sherry Rowland was well thought of and possessed of an excellent reputation, and young Dr.

Molina was on his way. But the two of them were about to enter a new realm of public and professional scrutiny, and he gently tried to prepare them, recommending they make their case as thoroughly and completely as they could, because they would be tested and questioned by the best, the brightest, and the most desperate of all: the chemical giants themselves.[13]

Rowland asked Johnston if he would like to join them in writing up the results—and entering the spotlight. Johnston grinned and declined. He had already taken his turn in the barrel.

The two chemists returned to the airport that evening with a last question from Harold Johnston ringing in their ears: "Are you ready for the heat?"

Rowland now faced a dilemma: an impending departure with his wife, Joan, for what was to have been a relaxing change of pace for five months in Vienna. He had planned to use the time to search for interesting new projects, but now he'd be leaving one of the most profoundly challenging projects of his career. Yet, in Europe he would have some good opportunities to discuss the theory with other scientists (having previously scheduled several scientific talks) while Mario continued to refine the data back home.

The Rowlands caught their flight as scheduled, but within three days of arrival in picturesque Vienna, Sherry Rowland was working on a paper to report the CFC–ozone connection, which he quickly submitted to the British journal *Nature*, mindful of the need to get the word out to an endangered world as quickly as possible. If they were right, he told Joan, "it looks like the end of the world!"[14]

It took five long months for *Nature* to get around to printing the Rowland-Molina paper, and for the two men whose names were on it, the wait was an agony of expectation and uncertainty. All hell was certainly going to break loose when the paper appeared, or so they figured. Harold Johnston's warnings had not been lost on them, and they spent much of the intervening time—Mario in the lab at Irvine and Sherry in Vienna—boning up on the atmosphere and anything else that would help them defend their case.

At long last the edition carrying the "Rowland-Molina Theory" appeared in the June 28, 1974, issue—to thunderous silence.

When "papers" are submitted to scientific journals, they are judged initially by the editors, but then sent out to respected scientists in the same field for a critical professional review of the worth of the work, its accuracy, and its relevance to society and science (not necessarily in that order). It had taken about four months for various referees to look over the Rowland-Molina paper and, with raised eyebrows, try to figure out what to do with the California scientists' radical new off-the-wall theory, and the connection between CFC's and ozone.[15] The reaction among the general scientific population seemed much the same at first as that of the media: suspended animation, followed by puzzlement, incredulous rereading, and the tendency to disregard a pure theory put forth by mere chemists who had no experience in atmospherics.

Several months before the paper finally came out, Sherry Rowland had been called by an upset public relations man from the chemical industry who had read about the "ludicrous" chlorine-ozone theory in a leaked Swedish newspaper article.[16] He had also heard that the Rowland-Molina paper was using the world "Freon" instead of chlorofluorocarbons. What, the man demanded, had Rowland been thinking of to slander a respected trademark by connecting it with doomsday theories based on nothing but theoretical laboratory chemistry? Rowland promised to use the generic term from then on, but he figured the complaint would just be the opening round from the industry. Yet during the summer following the paper's appearance there were only the sounds of silence from the CFC producers, and by August, Rowland and Molina started to wonder whether the scientific community in general had even noticed.

They had. But it was the galvanizing realization about

the delayed effect of CFC releases that froze many readers in their tracks with incredulous disbelief. How could *that* be true? Yet...

Almost all the chlorofluorocarbons that had ever been released into the lower atmosphere, according to Rowland and Molina, were still *in* the lower atmosphere, and going up. In fact, all of the stuff would eventually reach the Stratosphere in a process requiring decades. It was quiet and unemotional language, but the paper clearly pointed out that if the free chlorine did in fact have the predicted catalytic effect on the ozone layer, the effects could be expected to continue—and increase—for many more decades because of the impossibility of removing the slowly rising army of CFC's from the Troposphere. In other words, we could put it in the air, but we couldn't recall the millions of tons of the resilient little molecules, and that meant that if CFC-borne chlorine atoms ever started destroying the ozone layer with industrial-strength seriousness, the effects would be unstoppable for perhaps fifty years, and that was true *even if all CFC production was halted immediately worldwide*! Those were sobering words, even to those who doubted the chemistry.

To the chemical industry, the Rowland-Molina theory was a shocking assault on what had always seemed the perfect product, and they weren't going to take it lying down. There were six major U.S. producers of CFC's by 1974, and Du Pont Chemical had been the first, entering the commercial production phase back in the thirties.[17] Even though CFC's accounted for less than 2 percent of giant Du Pont's profits by 1974, the CFC lines were star performers, and as one scientist warned Sherry Rowland early on, Du Pont, for one, wasn't about to stop making them.

Du Pont, which prides itself on its reputation as a highly responsible company, had been careful about CFC's from the beginning, paying very close attention to any questions of CFC safety (if for no other reason than their suspicion that no product could possibly be so utterly safe and perfect).[18] In 1972 Du Pont had sponsored a two-year study on the overall safety of CFC's—a study that basically seemed to center on the Los Angeles basin. Three research groups were involved: one that tested CFC's in a smog chamber and concluded that CFC's would not react with L.A. smog; a second that measured UV interaction for CFC's (but stopped right in the middle of the stratospherically important UV-C range); and another that simply continued to measure CFC's in the atmosphere. The groups involved did not ask the same questions as Rowland and Molina, nor did they follow the same logic tree Mario Molina would later use, chasing the durable CFC's all over the Troposphere in calculations. Whether the effort was a serious attempt to find out if CFC's were dangerous or not, it was in the end completely ineffective.

Rowland and Molina had simply decided to look a bit further—or more precisely, a bit higher. The industry effort (orchestrated by Du Pont with what it described as the best of intentions) stopped at the Tropopause, and failed to consider the Stratosphere: the ultimate CFC dumping ground.[19]

During the summer of 1974, while Sherry Rowland and Mario Molina were waiting for the other shoe to drop, the two University of Michigan scientists who had worried about the space shuttle injecting reactive chlorine into the Stratosphere finally caught up with the Irvine chemists at a scientific meeting in San Diego. Ralph Cicerone and Richard Stolarski had read a preprint of the *Nature* article and were convinced the threat was real and immense.[20] The problem, though, was how to convince anyone else, especially policymakers who could take action. The conference ended with a small, local newspaper article about the worrisome Rowland-Molina findings, but the article failed to gain even the attention of the wire services. They would have to wait, apparently, for the publication in *Nature*.

Stolarski and Cicerone had no thoughts of getting themselves in the trenches in a shooting scientific war when they began poking into the Rowland-Molina hypothesis. They had simply thought that the questions of stratospheric ozone and shuttle-delivered chlorine were an interesting way of getting involved in atmospheric matters. They were neither advocates nor radicals. They were simply responsible scientists who were finding laboratory support for Rowland and Molina's conclusions, and in the course of normal scientific research, Cicerone and his University of Michigan group wrote up those supportive results and submitted them to *Science*, which scheduled the paper for a September 27 publication date. The magazine prepared a press release to go along with the paper, but even that wasn't to be used until the formal publication date.[21]

It was a classic dilemma for all four men. There was a moral imperative, and an ethical imperative, to speak out. Publishing probably wasn't going to be enough, but publication was the only formally accepted method for scientists. Yet, chlorofluorocarbons were so deeply ingrained in modern life—and especially in modern *American* life—anything less than major public attention focused on the problem would be insufficient to generate pressure on legislators.

In September, Sherry Rowland flew to Atlantic City to give the first major presentation on the issue to the monolithic and diverse American Chemical Society.[22] Maybe the paper hadn't yet stirred up a hornet's nest, but surely the ACS meeting—and any attendant press interest—might just get the necessary attention.

The "cause," however, was becoming more urgent than ever. Mario Molina had been refining the estimates of worldwide ozone loss, and they were becoming apocalyptic. The figures—based on a 10 percent per year increase in CFC production through 1990—predicted a 5 to 7 percent ozone loss by 1995, and a 30 to 50 percent loss by the year 2050![23]

That was galvanizing and controversial enough, but when Rowland tried to explain them in more human terms, he succeeded in ripping the lid off the Pandora's box that Harold Johnston had cautioned him would lead to excruciating "heat." A 30 to 50 percent ozone loss would mean a drastic rise in skin cancers, possible shift of climatic patterns because of less ozone to warm the Stratosphere, and possible crop damage. The lesson, said Sherry Rowland, was that what goes up has direct and damaging consequences, and CFC's, plain and simple, should be banned.

Banned! An eight-billion-dollar-per-year industry employing literally hundreds of thousands of Americans! That was unthinkable. Where was the evidence? This was *laboratory chemistry,* for crying out loud, with not a molecule of observational proof from the real world and the real atmosphere. Within the chemical community, at least, the presentation touched a blowtorch to some sensitive rear ends.

The media response was tepid at first. A few articles followed a press conference set up against industry wishes by ACS news manager Dorothy Smith, and their headlines were appropriately alarming, but as the weeks went by, more and more articles began to pay serious attention to the new theory, and what it might mean.[24]

Stolarski and Cicerone's research group, responding to the wire service coverage of the ACS news conference, announced their findings the next day by press release (their supportive paper was published in the American journal *Science* on September 27), giving Rowland and Molina substantial credibility and boosting the estimate of total ozone loss worldwide to 10 percent by 1985 to 1990!

Not to be outdone, Harvard atmospheric scientists Steven Wofsy and Mike McElroy (a flamboyant and combative scientist who had originally derided aspects of Stolarski's theories in Japan before the Rowland-Molina paper came to light) weighed in with ozone loss figures in the same general range in late September, gaining substantial national publicity in the process from the dean of American science reporters, Walter Sullivan of *The New York Times* (Sullivan mentioned Rowland and Molina only in passing).[25]

In England, the brilliant and fiercely independent atmospheric chemist James Lovelock (who devised the theory of

Gaia, which considers Earth a living organism with self-correcting and regulating abilities) was unimpressed. Lovelock had been the first scientist to actually detect the presence of CFC's in the atmosphere, and had proclaimed them of "no conceivable hazard," a position he began to rethink ever so slightly after reading the Rowland-Molina paper. "The Americans," he told a British newspaper (as reported by Lydia Dotto and Harold Schiff in their excellent 1978 book on the subject, *The Ozone Wars*), "tend to get in a wonderful state of panic over things like this. I respect Professor Rowland as a chemist, but I wish he wouldn't act like a missionary ... I think we need a bit of British caution on this."

Lovelock, disregarding his own advice about being cautious with one's comments, went on to say that the Rowland-Molina controversy was "like the great panic over methyl mercury in fish. The Americans banned tuna fish and they blamed industry until someone went to a museum and found a tuna fish from the last century with the same amount of methyl mercury [occurring naturally] in it." (James Lovelock could be forgiven for not knowing the name of that "someone" scientist who thought to go find a museum fish was none other than Sherry Rowland.)

Despite Lovelock's reservations, and despite the rising volume of rumbled discontent and alarm from the chemical industry, the Harvard and Michigan papers (along with several other supporting findings) gave the Rowland-Molina theory enough momentum to spur some significant action. On October 8 the National Academy of Sciences decided to name Sherry Rowland, Harold Johnston, Mike McElroy, and two others to a committee to determine the seriousness of the ozone–CFC threat and to decide whether a full-scale investigation was warranted. As the panel prepared to meet, the CFC industry rushed to prepare its case against the wild, unproven, highly suspect hypothetical Rowland-Molina theory which, in their view, threatened public panic and the sales of aerosols. On the latter point they were right: By late fall of 1974 the controversy would be blamed for dropping the sales of CFC–propelled aerosols by 7 percent. The war was on, and Rowland-Molina-Cicerone-Stolarski-Wofsy-McElroy and anyone else who had the un-American temerity to suggest CFC's might be dangerous to mankind were clearly the enemy.

While the University of California at Irvine stood resolutely behind Sherry Rowland and Mario Molina and never wavered, at the University of Michigan, Ralph Cicerone began feeling the heat of official discontent from his academic superiors, while on the far side of the issue, a wide range of environmentalists from the rather sedate Sierra Club to more wild-eyed and less responsible versions rose almost as a single body to get behind the Rowland-Molina theory and couple it to the broader goals of cleaner air and water. To the environmentalists, the fact that CFC's were used in spray cans full of noncritical convenience items provided a marvelous opportunity to attack the conscience of the American consumer right in his own bathroom (which on average contained at least three cans of ozone-killing aerosol products).[26] The oversimplified cry that Americans were trading the ozone layer for such unnecessary, frivolous uses of CFC's as aerosol deodorants and hair sprays was an effective argument in many ways, but it instantly obscured two major facts about chlorofluorocarbons: Although two thirds of American usage was through spray cans, CFC's were also used in the United States as refrigerants, and as foam-blowing agents (and to a minor extent as electronics solvents); and that any truly effective steps to protect the world's ozone layer would have to be on an *international* level, since the atmosphere above the United States was not walled off from the rest of the planet. Within months, however, the prime target in the public's collective mind had become spray cans and their CFC–propelled contents.[27]

The industry had only begun to fight. Even as Sherry Rowland and Mario Molina began to breathe a sigh of relief that there was a serious show of support from other respected scientists, buttressed by environmental outcries and increasing publicity, a multimillion-dollar counterattack got under way in the form of rapidly funded research projects designed to disprove the Rowland-Molina theory.

Du Pont, the company with the most to lose, led the attack, setting up truth squads of company spokesmen to refute the Rowland-Molina theory wherever it raised its ugly head—including scientific conferences. The theme was simple and somewhat compelling: American businesses generated over eight billion dollars of revenue from the manufacture and sale of CFC's and related products (such as the spray cans themselves), and the Rowland-Molina theory was merely a purist hypothesis that as yet had no empirical proof to back it up. Since the theory could be wrong, where was the logic in wrecking such a profitable industry employing so many Americans? More research, went their battle cry, is required before we damage such an industry.

But, came a quieter retort, what if *you're* wrong, and Rowland and Molina are right? Can we afford to wait for more proof and gamble with Earth's welfare?

And, of course, an even quieter corollary was raised: Should *any* scientific matter involving potential environmental harm of significant magnitude be decided by reference to the economics and profits involved?

Neither Sherry Rowland nor Mario Molina had ever sat down and discussed the question of just how far they should go in presenting their case to their scientific brethren, the politicians, or the press. It was simply assumed between them that they would do whatever was necessary to get the word out.

Neither of them had ever appeared before a congressional hearing, but both Rowland and Molina got their first opportunity before the end of the year. On December 11, 1974, accompanied by Ralph Cicerone, they testified before the House Subcommittee on Public Health and the Environment, explaining both the scientific case, and the reality that it was really up to policymakers to make a rational judgment based on what they already had before them. There was, Rowland told them, sufficient scientific evidence to justify at least *some* national action to limit, cap, or reverse the rapid growth in CFC production. After all, as their paper had shown, even an immediate worldwide halt to all CFC production would not solve the problem, because millions of tons had already been turned loose in the lower atmosphere. If CFC–borne chlorine really could harm the ozone layer, what had already been released into the Troposphere would keep the process going for at least a half century. Permitting industry to add to the problem while waiting for scientific confirmation was very dangerous, he told them, because of a simple and frightening delayed-reaction reality: By the time enough of the CFC's rose to the Stratosphere, broke down, and released enough free chlorine to cause a measurable decrease in the Earth's total volume of ozone, it would be far too late to prevent the effects from continuing for fifty to one hundred years. Our grandchildren, in other words, would live through the worst of *what had already been done!*

In the same hearing, Du Pont's Freon division manager, Raymond McCarthy, disregarded the essence of the warning Sherry Rowland was trying to deliver and told the panel that there was no reason for regulations until the theory had been proven. In his, and other witnesses' estimation, there *would* be sufficient time to stop dumping CFC's into the atmosphere if the Rowland-Molina theory were really correct. The fact that the beneficial effects of stopping CFC pollution would be roughly forty to fifty years in coming had been lost on almost everyone.

The Rowland-Molina theory, in the eyes of industry, was speculative and unproven, but Ray McCarthy did add that if it was ever proven that chlorofluorocarbons were a threat to ozone, Du Pont would stop making Freon. That promise, repeated later by the president of the company, would come back to haunt Du Pont.

Sherry Rowland flew back to California leaving behind

him the beginnings of a major battle of conflicting claims and opinions that were already leaving policymakers in confusion. Any layman who had listened to the testimony at the hearing would have come away unsure what to believe, and that inherent confusion was going to get far worse much faster than anyone suspected.

The few lawmakers who had already entered the controversy were indeed feeling confused. Here they were being advised by trustworthy and sober scientists on one hand to take immediate action to protect the ozone layer based on a set of findings—a theory—that had not been disproven, yet a theory that others derided as premature or worse. In the same hearing they would be told that the Rowland-Molina hypothesis was either correct, compelling, and urgent, or unproven, speculative, and highly suspect. There was no way for the average congressman, senator, or staff member to have any earthly idea which conclusion was the correct one. And, overriding any environmental-protective instinct was the dark warning that premature legislation could throw hundreds of thousands of American voters and constituents out of work, all on the basis of a theory that could be proven wrong at any time. What was worse, the scientists themselves couldn't seem to agree even when they were trying to support one another's position! As one deeply concerned staff member put it, "I'm waiting for these scientists to get their act together and make a clear decision. Is the sky falling or isn't it? If *they* don't know, how the hell can *we* know? I mean, even Chicken Little and Henny Penny had clear and certain opinions. These guys keep changing theirs."

But that was the crux of the dilemma, because the CFC–ozone issue ignited by Sherry Rowland and Mario Molina was not unique in presenting policymakers with rampant, scientific uncertainty. The consistent lesson is not that scientists have difficulty agreeing and being certain, because that will always be the case. The lesson is that the policymakers have difficulty understanding the timing: There comes a point where a value-judgment-based policy decision has to be made based on the best data available at that moment, regardless of continued scientific arguments. The greater the potential risk to society, the quicker that decision point must be reached. The scientific community can tell all they know and even recommend what actions might help, but only the policymakers themselves hold the key. Throwing that responsibility back in the face of the scientific community is a dangerous avoidance of responsibility.

The year 1975 dawned on exactly that dilemma in Washington, D.C.

If the scientific community seems to be having trouble coming to a scientific conclusion, one tried-and-true solution is to appoint a board full of other scientists and force *them* to pass judgment. Even if that fails, it gets the monkey off the back of the politicians for a while, and allows them to avoid making what might be a controversial decision. In early 1975 the Ford administration used that very methodology when the president appointed the Committee on the Inadvertent Modification of the Atmosphere (IMOS). Of course, formation of IMOS did prove that the White House was aware of the problem and actively doing something, but that was little comfort to scientists such as Rowland, Molina, and Cicerone, who felt the severity of the threat to the ozone layer was so grave, the mere *possibility* their theories were right would justify a ban on at least nonessential uses of CFC's. What was needed was a value-judgment-based policy decision, but creation of IMOS threatened to throw the issue right back in the scientific arena as a "science court" test of the Rowland-Molina theory.[28]

IMOS decided in June 1975 that restrictions might indeed be necessary (a value-based policy decision), but first someone would have to decide whether chlorofluorocarbons were, in fact, a hazard to the ozone layer (a scientific decision). The initial verdict, in other words, was that the scientific community—and in particular the Rowland-Molina theory—must face a science test of sorts, and for that role, IMOS tapped the federally chartered National Academy of Sciences (NAS) on the shoulder and dumped the entire mess in their laps. If *they* determined that the Rowland-Molina theory was valid, said IMOS, then the government *should* act to restrict the production and sale of CFC's. At NAS, the assignment was about as welcome as a wet and muddy dog in a spotless living room.

The prestigious NAS tries hard to maintain as much balance and independence as possible, keeping the greatest possible distance from political maelstroms. Suddenly, however, it was in the midst of a political tornado, which was not a comfortable feeling. Anything they did might be wrong and hazardous to their reputation and the structure of governmental support.

The first NAS response, quite logically, was to stall for time. Five months passed before the first committee was even named to tackle the problem, and over a year would pass before the members finally gritted their teeth and announced something loosely resembling a conclusion. Considering their previous involvement with ozone, the discomfort was understandable.

NAS had not compiled the most trustworthy record when it came to being the supreme court of American science, and Dr. Harold Johnston of Berkeley in particular was not likely to forget the savaging both he and the Academy had taken at each other's hands in 1971. The NAS had stood behind a rather poorly supported earlier conclusion that SST's would have no effect on the ozone, and when Harold Johnston spoke out on the other side of the issue, his findings and views were treated with skepticism bordering on contempt. NAS members tended to regard Johnston as an overly emotional advocate, and he had come to regard them as gullible, ill-informed putty in the hands of the political considerations of the sitting administration.

In the Rowland-Molina crisis, however, NAS finally named a twelve-person scientific panel and divided it into two subgroups, one to rule on the validity of the Rowland-Molina hypothesis, the other to tell government what should be done about it. While Sherry Rowland, Ralph Cicerone, and Mario Molina were all fairly confident of the balance and fairness of the panel and the process, they all knew the deliberations would be made in anything but a vacuum. The chemical industry in general, and the CFC producers in particular, had plenty of friends in high places, lobbyists, and resources, and they sure as hell weren't going to let the NAS decision-making process proceed without their input, especially since legislation had been introduced in both the House and Senate that would amend the Clean Air Act and ban CFC's in one degree or another *if* the NAS panel found them hazardous. Whether the Academy wanted it that way or not, they had in effect been appointed a supreme court of science, though to extremists on both sides there was fear that what IMOS had in fact created was a star chamber.

While the NAS process got under way, various hearings and staff meetings on Capitol Hill proceeded as well, belying the fact that almost none of the staff members working for the various lawmakers were scientifically trained or knowledgeable in the area of chemistry or atmospherics. With industry snorting about this "unproven theory" and painting dire pictures of hundreds of thousands of chemical workers and their families put out of work, it seemed the CFC producers were pulling together, which meant there was a desperate need for a united front on the Rowland-Molina side of the equation. Other than the accord among Sherry Rowland, Mario Molina, Ralph Cicerone, and a few others, however, it was business as usual at the OK Corral of science, with competition, backbiting, second-guessing, and opportunism ruling all too often. Lawmakers would listen to a familiar drumbeat of caution against hasty action in the face of an unproven theory that had no empirical support, and then an array of chemists and other scientists would march across the dais emphasizing different reactions and discussing different predictions, some higher, some lower, and some not even affecting the same debate! The nitpicking disagreements simply took what was already a perplexing issue for the layman and made it impossibly confusing.

"Exactly what the hell *are* they saying? Are we losing ozone or aren't we?"

Even Mike McElroy of Harvard, who had originally followed Steve Wofsy in early support of Rowland and Molina, watered down his position to such a degree that his testimony in one hearing served to directly damage the credibility of any ozone-chlorine-CFC connection, especially Sherry Rowland's.

And over and over, the fact that there were no direct atmospheric measurements of ozone loss flew in the face of the chilling reality that no actual ozone loss might be measurable for quite a few years, but if CFC production wasn't stopped, by the time a loss was seen it would be too late to prevent a catastrophe. Mario Molina knew it. Sherry Rowland knew it. But that point kept getting lost in the static of competing nitpicking and scientific fine tuning.

Despite the intimidating appearance, the industry was not a united front, and on at least one occasion Du Pont specifically disassociated itself from some of the more hysterical industry-funded attacks on the Rowland-Molina theory. In mid-1975 the CFC producers were stunned when the fifth largest producers of CFC–powered aerosols, Johnson's Wax, announced they were ending production of CFC–filled products. It seemed a dark betrayal, yet the ranks of those who saw the impending regulations coming and decided to switch rather than fight would begin to grow in the following year.

Rowland and Molina were spending more and more of their lives in hotels and hearing rooms, and CFC's were still being produced at significant rates (though sales of aerosols were dropping rapidly). Over two years had passed since the "midnight discovery," yet the fight showed no signs of ending, and they were forever, it seemed, defending their theory and trying to get people to understand what it meant at the federal level.

At the grass roots level, though, things had begun to happen more rapidly, and movement toward legislation to ban CFC aerosols began in several state legislatures, including that of the state of Oregon. Though some chemical industry spokesmen derided the efforts as "stupid" and "essentially silly" ("What're they going to do, put a plastic dome over their state?"), industry "truth squads" burned up considerable sums of industry money in attending every state legislative hearing (and even one city council hearing) to counter the Rowland-Molina theory.

Oregon, however, fell first, banning CFC aerosols in June 1975 and providing a stiff penalty for selling such ozone-destroying products. The governor of Oregon, in signing it, acknowledged that the action was based on unproven theory, but that it was wiser to "err on the side of caution."

In Washington, D.C. the months continued to tick by as the NAS panel agonized over its task. Sherry Rowland was getting genuinely frustrated with the wait, however, and all the more so when he heard that the Department of Commerce had already come out against CFC regulations even before the NAS report was finished. Equally worrisome was a statement by the Environmental Protection Agency's chief calling for an international solution. Certainly worldwide CFC regulation was vital, but for the EPA to call for it at that moment would give the industry a new rallying point: Why should we act unilaterally? No federal CFC regulations until the rest of the world joins us in a treaty!

No matter who was doing the arguing, the heart of literally every debate concerning CFC's held exactly the same key question: Was the Rowland-Molina theory right or wrong, true or false, operative or inoperative? In the middle of the maelstrom of scientific, industry, political, media, and public charges, countercharges, claims, outcries, apathy, and scary predictions stood the same two affable, highly intelligent, quiet and conservative chemists. Somehow, Sherry Rowland was keeping himself calm, and Mario Molina was spending as much time as possible in research, but if the winds blew gently around them, it was only because they stood in the eye of the hurricane. It was not the safest of all possible positions to occupy.

By midsummer of 1975 researchers at the National Center for Atmospheric Research (NCAR) and the National Oceanic and Atmospheric Administration (NOAA), both in Boulder, Colorado, had become fully involved in helping NAS make their decision by searching for actual, measurable readings that could shed some light on "the Theory." Three steps were considered necessary to prove the Rowland-Molina findings. The first was to show that CFC's were actually reaching the Stratosphere in their original molecular shape, and the second was proving that ultraviolet light breaks up the CFC molecules into their various components, releasing chlorine or reactive chlorine compounds (such as chlorine oxide). Both had been accomplished by early fall. The last step—and perhaps the most important—was the search for the chemical "smoking gun": ClO, or chlorine monoxide. There could be no ClO in the Stratosphere unless it was the by-product of ozone destruction. If ClO wasn't there, Rowland-Molina were wrong.

Out in east Texas working with a limited budget in a far corner of atmospheric science at a NAS facility called the National Scientific Balloon Facility, another tall, congenial scientist named Jim Anderson was busy trying to answer question number three with eight-hundred-foot-long balloons made of thin polyethylene film and carrying huge packages of sensitive instruments to altitudes of 150,000 feet. Anderson, a University of Michigan research associate, who with his prematurely white hair and broad, craggy face resembles actor Peter Graves, in effect had before him a scientific version of an impossible mission: finding evidence of a particular molecule scattered in the atmosphere at concentrations of no more than two parts in a billion.

Before Jim Anderson found the smoking gun, however, the Rowland-Molina theory was nearly blown out of the sky by another sudden "midnight discovery" back in California. And the scientist who tripped over the new evidence? Mario Molina himself.

In the societal scheme of things, lawyers are never supposed to lie to judges, reveal privileged client statements, or play around with the funds clients entrust to them. Those are bedrock ethical requirements, similar to a doctor's Hippocratic oath. In science, there is a traditional ethic that scientists must never hide or understate known exceptions and limitations to their theories. Scientists are never to conceal new, contradictory findings, no matter how destructive they might be to theory or reputation. Science is, after all, the search for the truth of things in the universe as we can know it, and truth, either as quality or as commodity, is always in limited supply. A very common theme stretches back through science even to the days of alchemy: Today's truth is tomorrow's disproven theorem. A scientist who discovers something new that assassinates yesterday's truth, but who fails to report it, is not a true scientist in the eyes of that calling. Thus there was no hesitation on the part of Mario Molina or Sherry Rowland in telling the National Academy of Sciences panel—which was deliberating on the veracity of their theory—that Mario had found something new in the lab that put the original estimates of eventual ozone loss in serious doubt.

Aided by Dr. John Spencer, a new chemist who had joined the team at Irvine, Mario had been checking methodically for any other molecules that might in any way become a sink for chlorine, or somehow interfere with the chlorine-ozone destructive catalytic cycle in the stratosphere. Mario had long been totally at home in the back stacks of dusty science libraries engaged in scholarly research, and it was in just such a foray that he found in some old German chemical literature some references to previously unexplored properties of a molecule he hadn't paid much attention to before: chlorine nitrate. He and Sherry Rowland had considered chlorine nitrate in 1974 as a possible sink for chlorine, but its lifetime would be too short to limit the chlorine's destructive attraction to ozone. Or so they thought.

Now, however, a series of delicate but straightforward lab experiments to validate that conclusion suggested something entirely different: The chlorine nitrate compound was much more stable than they had believed, which meant that it *could* tie up large numbers of chlorine atoms.[29] But when Mario cranked the new figures into his mathematical models, the total amount of projected global ozone loss they had originally pegged at 7 to 14 percent plunged through the floor! (In fact, in one set of calculations done elsewhere, it looked for a while as if the net result of added Stratospheric chlorine could be an ozone *gain*!) Molina and Rowland were in shock, and a bit of agony. They had to put the world on scientific red alert, perhaps for nothing.

To make matters more painful, the NAS report, which had been awaited with great trepidation by both of them for so long, was finally scheduled for release in a matter of weeks, and the new data would undoubtedly send everyone back to the lab, costing many more weeks.

Nevertheless, an immediate phone call went out to the NAS group from Sherry Rowland, reporting the results. Predictably, the NAS panel was thrown into immediate confusion, and late-night sessions began in numerous labs and computer-modeling facilities to try to find the true extent of the role of chlorine nitrate in tying up free chlorine in the Stratosphere.

Within days, other labs and chemists had refined the models to show that ozone loss would occur and would be significant, but the chlorine nitrate would to some degree become a sink for chlorine that had come originally from CFC's, lessening the overall effect on the ozone layer of the Earth. So they were right, but not quite as right as before (there certainly was no production of ozone—that finding was wrong). The entire affair could have ended there if the issue was not also an economic one, but as Sherry Rowland feared, the CFC industry pounced on the news like a duck on a June bug and ran to the press trumpeting that the Rowland-Molina theory had been disproved.

It had *not* been disproved. It *had* been ameliorated, at least for a while. It took several months and some very long hours slaving over hot computers and calculators before scientists such as Paul Crutzen working with Ralph Cicerone were able to report back to the NAS panel that the chlorine nitrate changed some aspects of the problem, but that the overall loss of ozone would still be toward the low end of the Rowland-Molina estimates of 7 to 13 percent worldwide. The flap delayed the NAS report by five months, but since the scientists who had brought forward the challenge to Rowland-Molina were none other than Rowland and Molina themselves, there was a beneficial effect: an added veneer of trustworthiness in the theory. If even the progenitors were still working on the problem and reporting honestly on a real-time basis, their neutrality was far more substantial than their opponents had wanted to believe. Now even the CFC industry grudgingly began to acknowledge behind closed doors that there was reason to believe in "the Theory" and the inevitability of CFC regulation.

On September 13, 1976, NAS released their report with the ruling that Rowland and Molina were correct. The Earth, they concluded, would in fact lose somewhere between 2 to 20 percent of its total ozone protection, with 7 percent being the most likely figure, if nothing was done to stem CFC production.

But the second section of the NAS panel had reached a different conclusion in effect. While it acknowledged the finding that Rowland and Molina were right, it recommended that no immediate regulation be enacted. The government, it said, should be given two more years to study the problem. The Academy, in other words, had attempted a compromise; in the end, all it had succeeded in compromising was the integrity of its effort.

Nevertheless, Sherry Rowland and Mario Molina felt vindicated, and within a month the U.S. Food and Drug Administration disregarded the "wait for more research" recommendation and moved to phase out nonessential CFC use in aerosols—a beginning victory that had taken three years of hard battles and untold personal and professional tolls on both men.

Things began to happen at a faster pace, and by early 1977 many different CFC–aerosol packagers began switching to pump dispensers for household products and advertising heavily to promote the "ozone-safe" qualities of their new, non–CFC lines. Such moves, of course, were in anticipation of the inevitable, which finally occurred on May 11, 1977, when several government agencies jointly announced a timetable for mandatory phaseout for nonessential CFC aerosol products, to be effective in late 1978.[30]

The battle—the first battle—had finally been won (though only in the United States, and soon after in Canada, Norway, and Sweden). But the war, so to speak, was just beginning. There were many other uses of CFC's continuing unabated, and millions of tons of the stuff would continue to be manufactured—and released—each year all over the planet. Even aerosol spray cans continued to be produced and sold outside the United States. Yet the American tendency toward pragmatism, and the desire to say, "Okay, that's done and over with, let's get on to the next problem," began to operate, and the war was considered won. Americans have a penchant for clean solutions and heroes riding into the sunset—a "good exit," in acting terms. But while Rowland, Molina, Cicerone—and many others who were now convinced of the threat—watched in utter dismay, the audience began stampeding out of the theater, falsely secure in the "knowledge" that the ozone-killing properties of CFC's had been eradicated by the ban on spray can usage. Even the fact that only four countries had instituted such a ban failed to register on the public. As the seventies came to a close, the perception took root that the ozone war was over, that there was no more ozone problem. That perception was aided by a string of unfortunate reports over the next few years that had ozone loss estimates fluctuating all over the map, from zero loss to greater than 20 percent.

NAS itself was responsible for much of the confusion, issuing well-meaning reports that in 1979 pegged projected ozone loss at a whopping 16.5 percent, but downsizing that to a 5 to 9 percent loss in a 1982 report, and a 2 to 4 percent loss in 1984. The only consensus during those years was that the atmosphere was a vastly more complex place chemically than anyone had figured, and although the basic chemistry of the Rowland-Molina hypothesis was never questioned, the total global ozone loss figures simply defied precise estimation. With the world concentrating on other things—and the election of President Ronald Reagan, to whom environmental movements were direct threats to free enterprise and the American way—Sherry Rowland was fighting a losing battle.

In May 1981, Reagan's new Environmental Protection Agency administrator, Anne M. Burford, testified before her Senate confirmation hearing inquisitors that, in her opinion, the CFC issue was "highly controversial." Indeed, in a book published long after she had been forced out of the EPA for doing practically nothing to protect the environment, she would reveal her true attitude toward all the scientific work that had gone into validating Rowland-Molina: "Remember," she would write, "a few years back when the big news was fluorocarbons that supposedly threatened the ozone layer?"[31] With such attitudes and continued waffling in the scientific estimates of ozone loss, the advice Sherry Rowland was getting was turning toward the "abandon ship" line of reasoning. He had done enough. He was wasting his career.

Mario Molina left U.C., Irvine, in 1982 to head up a research group at the Jet Propulsion Laboratory in Pasadena, leaving Rowland still engaged in refining the ozone research. Though he had become an assistant professor in 1975 and no longer reported to Sherry Rowland, Mario Molina continued to collaborate with his former boss, and stayed in effect at Rowland's side throughout the years of controversy over what had become known worldwide as the "Rowland-Molina theory." Rowland, however, had taken most of the heat. Being more comfortable in the spotlight of public scrutiny than Molina, Rowland had become the visible element of the team, and the fallout from that exposure was taking its toll. No longer considered a "balanced, neutral" scientist, Rowland received no invitations to speak at chemical industry functions, and very seldom was invited to speak at other university chemistry departments. Such an institutional snub would have been very puzzling for a senior chemist of his stature— except for his advocacy of "the Theory." There was a penalty to be extracted for those who, regardless of motivation, care, responsibility, or correctness, have the unmitigated temerity to take science into the arena of policy or (shudder) policy recommendations. While Ralph Cicerone felt only minor setbacks in his career and Mario Molina even less, Sherry Rowland's professional horizons without question were affected. It was a form of negative feedback that his students—and many others in graduate and postgraduate programs—saw and took to heart. Here is what happens when you become "controversial." Take heed, young scientist, and keep your mouth shut.

Dr. Susan Solomon, a former doctoral student of Harold Johnston at Berkeley who had joined NOAA at Boulder, knew Sherry Rowland and held him in the highest esteem personally and professionally, but in a meeting in 1983 she advised the senior chemist to give it up: "The projected ozone losses are too uncertain and low—in the noise level. You're wasting your time, Sherry."

It was a warning she was to remember acutely later, in the spring of 1985, when a certain incendiary scientific paper landed on her desk for review.

CHAPTER
3
The Invisible Hole in the Southern Sky

Boulder, Colorado, March 1985

Susan Solomon came forward in her office chair, her coffee cup poised in midair as she reread the opening abstract of the scientific paper before her.
WHAT?
The word was formed but unspoken, yet it echoed through her mind as she turned the page for a moment to examine the various figures and graphs, returning then to the opening text. My God, she thought, if this data is valid ...
The peace and quiet of a Sunday afternoon filled the rather ordinary hallways of Aeronomy building with the delicious potential for reflective thought and uninterrupted reading, and Susan had developed the habit of migrating to her office on such days to catch up on various projects—as well as just think.

Sometimes she'd spend the afternoon writing—plowing through the reams of paperwork, letters, scientific papers, and other such tasks perpetually waiting for her attention. And sometimes she'd just luxuriate in the silence of the place and read. There were always new papers and journals filled with more of the scientific literature that forms the lifeblood of science—literature that flows through countless offices, outposts, and laboratories the world over as a written river of ideas, findings, postulations, and prognostications carried relentlessly and endlessly to thousands of highly trained minds, the yeast of intellectual ferment. For a scientist, keeping up with "the literature" is like breathing, a natural and essential function. And for highly accomplished scientists of growing stature in their selected fields—Ph.D.s such as Susan Solomon—there were sometimes added duties, such as reviewing papers for various professional journals to help determine whether they were of sufficient quality and scientific value to justify publication (there are usually far more papers looking for space in the journals than the other way around, and editors have to pick and choose).

A position such as hers in the Aeronomy lab of the National Oceanic and Atmospheric Administration's Boulder campus didn't require her to do prepublication reviews, but then no one had to assign anything to Susan Solomon. She was well known as an energetic self-starter, a chemist of the first order who had become fascinated with atmospheric chemistry and computer modeling and who was making significant contributions in those areas. Harold Johnston of Berkeley had been her Ph.D. adviser, but she had been lucky enough to get a fellowship to study at the National Center for Atmospheric Research, and learn under the guidance of its director, Paul Crutzen. Being accepted as a fresh "postdoc" at NOAA had also been an unusual coup, and she had thrived in that environment, probing with great energy the various projects before the Aeronomy lab, attending scientific meetings all over the world, and thoroughly enjoying the practice of quiet scientific research. Then, without warning, while in her senior year at the Illinois Institute of Technology, her attention turned skyward. The native Chicagoan had been involved in some routine lab measurements when she realized that the type of molecule she was studying had been found to be important to the atmosphere of Jupiter. Susan had been looking no further than the maze of glass and metal before her, but *Jupiter*! Suddenly her task became far more exciting than she had imagined, and the scope of what she was investigating seemed to soar quite literally into space. Wow, she had thought, chemistry *is* more than just test tubes! She had been hooked, then, on atmospheric chemistry, and now at the still-tender age of twenty-nine, Dr. Susan Solomon had amassed a list of credits and accomplishments usually reserved for someone five to ten years her senior, a fact that did not escape the notice of peers or outsiders.

"My age," she had once told an audience during a speech, "becomes less of a problem each year."

The background hum of the building's heating system fighting the chill of a Colorado winter day went unnoticed as she turned another page, devouring the paper with an increasing volume of questions stacking up in her head.

The paper had originally been sent to another scientist in Boulder, but he had been too busy and suggested Susan for the job, and she had agreed. It had been a routine decision and a routine function—until now.

The coffee cup was back on the desk, half-filled and unattended, as Susan went over the last line and returned to the first page, rereading from the top. It was her habit to read a paper several times, but in this case it would be vital. If the authors had found what they obviously were convinced they had found, it would be one of the most significant—and threatening—findings of the decade. She wondered who else had seen the preprint? It was like holding the first Teletype news copy of some amazing event.

Susan looked at the author's name again. The paper was

being considered for the British journal *Nature,* and the lead author was unknown to her, an atmospheric chemist by the name of J. C. Farman, of the British Antarctic Survey.

Whoever he was, by the end of the afternoon he had her vote—her enthusiastic recommendation to *Nature* that the paper should be published. She could find nothing inherently wrong with it, though some of Farman's suggestions about what might be causing this . . . "hole" in the ozone layer—for want of a better term—sounded a bit suspect.

She had no reason to question the accuracy of his data. Farman and his coauthors, B. G. Gardiner and J. D. Shanklin, had carefully addressed the first questions they knew a reader would ask: Were the instruments calibrated? Were the difficult-to-handle Dobson instruments properly operated?[1]

The answers were yes. The low readings in Antarctica did not appear to be due to poor observations or badly calibrated instruments. The readings were real.

> . . . within the bounds quoted, the annual variation of total O_2 at Halley Bay has undergone dramatic change.

That was probably going to prove an understatement.

Reviewers normally return preprint reviews to the authors without a signature, ducking the embarrassment of giving a negative review or potentially irritating suggestion to a colleague by remaining anonymous. Susan, however, signed her review and dispatched it back to the United Kingdom. Yet the paper stayed in her head, and over the next few days began taking more and more of her conscious attention. How could that tremendous ozone loss be documented? And was it the result of CFC–borne chlorine at last?

One of the first questions was why such low figures could be showing up above Halley Bay, but nowhere else. There had been no other reports of such a drop, as far as she had heard. And if the loss was real, where was the supporting satellite data? NASA had the ability to measure the total ozone all over the southern polar region from space, and such a hole should show up in their reports. It impressed her that Joe Farman's baseline—the length of time his team had been tracking Antarctic ozone with yearly figures—spanned twenty-seven years. Nevertheless, someone else *must* have noticed such a "small" detail as a loss of over 40 percent of the ozone layer, even over the most remote and forbidding spot on Earth.

Susan began digging through the literature looking for other ozone measurements from Antarctica. Joe Farman had tried to explain the massive ozone loss in his paper, but at least one part of his explanation bothered her. He suspected that the very low temperatures in the Stratosphere of the frigid Antarctic night permitted chlorine to destroy ozone when the sunlight returned in September.[2] If so, it might be the first major atmospheric validation of the Rowland-Molina theory, though the magnitude of the ozone loss was beyond Sherry Rowland and Mario Molina's worst nightmares. Ozone wasn't just being diminished, it was disappearing during the Austral spring when the perpetual darkness of the Antarctic winter was ending. A 40 percent reduction was almost beyond belief!

Susan knew that the chain of chemical reactions among what were essentially the various molecules in a gas ("gas phase kinetics" is the technical term), which Joe Farman thought might be the explanation, would indeed play a significant role around thirty kilometers above the Antarctic continent. Yet what bothered her was the fact that most of the ozone above Antarctica was *below* thirty kilometers, and Farman's solution wouldn't work as well there. Of course, measuring ozone with a Dobson Spectrophotometer from the ground meant you couldn't know what altitudes held what concentrations. All Farman's people could do was measure what was called the "total column," sort of a vertical core sample of the atmosphere, with all the findings averaged from low to high altitude, and Farman had certainly pointed out how much he needed better measurements to make a better guess.

No, she thought, he had some very sound points to go along with the main course—the shocking numbers of ozone loss—but there was something more that needed to be worked out. Now what the hell *was* it?

How, for instance, could such reactions occur in *gas phase* that fast? After all, the ozone that had been lost had disappeared within *one month,* yet the sort of catalytic reaction envisioned by Rowland-Molina would require several years to complete such a massive ozone-eating operation. Molecules floating around in the atmosphere in certain concentrations bump into each other with rather predictable frequencies, but not that fast. It seemed unlikely that free chlorine—no matter how energetic a sundance it did when sunlight returned to the Antarctic sky—couldn't possibly gobble ozone at the rates Farman's people had documented. The observed rates were more like the sort of reactions you'd get in a liquid state where the molecules were already rubbing elbows and didn't require random collisions to react. But this was a gaseous environment Joe Farman had been observing. Comparing the frequency of collisions between molecules in a liquid with the frequency with which they bump into each other in a gas is akin to comparing the number of potential fender benders during rush hour on the Pasadena Freeway to that of a Sunday afternoon drive down a deserted superhighway in the middle of Nevada.

And the **Stratosphere** is in a gaseous state—more like the Nevada deserts from a lowly molecule's point of view. It can get kind of lonely where bumping into others is an infrequent pastime.

Whatever was happening in chemical terms, the size of the ozone loss simply had to have been spotted by another team of observers *somewhere,* and sure enough, Susan finally located another set of Antarctic observations in a paper by Japanese atmospheric chemist S. Chubachi published in 1984. In 1982 Chubachi's team had measured a different parameter (ozone-mixing ratios) than Farman's team, and had done so from *their* research station at Syowa, Antarctica. Their data, though they hadn't realized it, showed losses similar to the Halley Bay data for 1982. Something catastrophic was obviously happening to ozone in both reports, and over both stations.

There was another unexpected benefit, though, that would end up being quite important in helping Susan and the rest of the atmospheric community piece together the puzzle: The Syowa data suggested that most of the "missing" ozone was nearly fifty millibars, which means about twenty kilometers above the ice—the place where most of the Antarctic ozone is as the long winter's night ends. There are several types of molecules commonly found at that level (such as hydrogen chloride, HCL, and $ClONO_2$) that can tie up chlorine, preventing it from destroying ozone too rapidly. The term for such a function is "reservoir," a molecule that forms by taking a chlorine atom or two and tying them up with such molecular stability that they simply aren't allowed to break out and go play with the ozone. Obviously, whatever reservoir molecules were floating around at twenty kilometers in the Antarctic morning were not making much of a dent in the voracious appetite of the chlorine, *if* chlorine was the culprit. That too needed confirmation. It was shaping up into the best of dilemmas and the worst of environmental nightmares: a major threat to the ozone layer cloaked in a scientific riddle that would be unbelievably fascinating to unravel.

Joe Farman's solution was a respectable beginning, but it would take a while to think of a more viable alternative.

Nearly twelve months, to be exact.

In the Antarctic spring a phenomenon occurs high in the Stratosphere that paints the sky with magnificent brushstrokes of color. The "canvas" for those beautiful displays are frozen ice crystals of water and other chemicals strewn across the lower Stratosphere, catching the sunlight at very low angles on the horizon. Rich reds set against indigo blue, brushed by the fire of orange, glow in iridescent profusion like mother-of-pearl, as if all the colors of the Aurora Australis—the southern equivalent of the Northern Lights—had been turned to shimmering ice and bathed in celestial spotlights. Polar Stratospheric Clouds form in temperatures on the south side of $-80°C$, air so cold that even tiny molecules of acids can freeze into ice crystals and join the beauty of the formations of frozen aerosols as they hang there some fifteen to twenty kilometers above Antarctica.

It so happens that fifteen to twenty kilometers is also the location of the winter Antarctic ozone layer.

Susan had yet to see a Polar Stratospheric Cloud (PSC) in person, but images of PSC's in photographs flashed through her mind as she compared the numbers and the possibilities over the next few months. Slowly, steadily, a hypothesis would begin to suggest itself through conversations and lively debate, a vague idea that would eventually take the form of a fascinating and previously unanswered question.

It wouldn't necessarily be that specific question that would change her life, but it was part of the process. And Joe Farman's paper had started it all.

Cambridge, England, April 1985

Joe Farman had come home in a state of excitement. Or perhaps it was wonder. His wife couldn't really tell as she regarded her husband at the end of the day in their hallway.

"My God, I can't believe it!"

"What, Joe?"

"My paper. It's being reviewed by a *woman!*"

She saw he had a grin on his face, and that was fortunate in Mrs. Farman's estimation. Joe Farman was no chauvinist—

as she knew, and Susan Solomon would later confirm when their professional friendship resulted in mutual visits. But there were precious few women in British scientific circles, and the gender revealed by Susan's signature had left him startled.

Her recommendation to publish, however, had come as a welcome relief. Farman's paper had already pulled in a very negative, unsigned review, and he was having trouble with the *Nature* editors over whether they would indeed print it. There was no way, according to the other reviewer, that the British data could be correct, and therefore no way the paper should be published.

Nature had received it on Christmas Eve, and here it was April, with precious time ticking by. The information contained in those pages was far too vital to wait any longer, so it was with substantial relief that Joe finally got the news that it would appear in May.

Joe Farman naturally assumed the paper would galvanize the scientific community, bursting over the heads of his peers around the world and creating an immediate stir. But with few exceptions, his assumption was wrong.

"Who the hell is Joe Farman? And who is this British Antarctic Survey, anyway?"

When the British paper finally appeared (in the May 16, 1985, edition), it debuted to roughly the same level of thundering silence that had greeted the Rowland-Molina paper in 1974. The figures from Halley Bay were simply too astounding, and no one seemed to know enough about Farman's group, or even the location of Halley Bay, to judge whether they did good work. The data—despite Farman's careful statements to the contrary—might still be the result of poor observations made by ham-handed technicians with deteriorating equipment looking at the wrong thing in the wrong place at the wrong time. Who knew? But a 40 percent ozone loss simply wasn't a very credible-sounding figure.

Not at first, at least.

In Pasadena, Mario Molina *did* feel like a bomb burst had gone off over his office at the Jet Propulsion Lab as soon as he had read the paper. He too had been unaware of Joe Farman (though Farman was quite aware of him and Sherry Rowland), and Mario had heard nothing whatever about the paper, or that any data regarding large Antarctic ozone losses even existed. That was very odd, he thought. Good data or bad, when there's a galvanizing new finding, most scientists in the field hear about it rapidly. A scientific grapevine catalyzed by perpetual meetings and laced together by a delicate but constant web of telephone calls and letters usually made the actual publication of a paper yesterday's news to someone whose work it directly affected. And perhaps no one on the planet was more affected than Mario Molina and his counterpart some miles to the south in Irvine, the equally startled Sherry Rowland.

"This must be chlorine chemistry at last!" was their common comment.

Rowland was getting ready for another of the never-ending treks to another scientific conference when the paper arrived. This trip was to Switzerland, and he bundled Xeroxed copies of the paper into his briefcase to test the reaction in Europe.[3]

In Cambridge, Massachusetts, K. K. Tung was as stunned as Mario Molina to find such a paper only after it was published. But if these data were reliable, something very unusual was going on in Antarctica, and it might involve as much the movement of the various air masses as it involved chemistry. A phenomenon called the "Antarctic vortex" was well known to atmospheric scientists and meteorologists alike, and it set up some strange conditions over the pole that were duplicated nowhere else on Earth. During the exceptionally cold temperatures of winter, the slow rotation of air around the South Pole formed a partially walled-off mass of air as large as the North American continent. What's more, the air within that vortex was kept away from the atmosphere outside for many months, becoming somewhat stagnant in a broad sense. Could *that* have something to do with the ozone loss the British were seeing? Tung, like Susan Solomon, was intrigued enough to start digging.

Through the first two months of summer Joe Farman's paper made the rounds of the atmospheric scientists, raising eyebrows but gaining few converts. Even Sherry Rowland was surprised to find the European response something on the order of a bored yawn. In the village of Les Diablerets near Geneva he had shown the paper to numerous other chemists, but finding someone who was truly excited over the Halley Bay discovery was not easy.

Even in the United States, the level of interest through August was minimal at best. There had been too much flip-flopping of ozone loss predictions in the National Academy of Sciences reports over the years, and there had been no verifiable worldwide trend of thinning ozone (though the 1982 NAS report did mention observed ozone losses in the upper stratosphere). The atmospheric science community, in other words, had been given time and reason to become almost numb to the issue.

Susan Solomon's pre–Halley Bay advice to Sherry Rowland to leave the CFC–ozone issue and delve into other projects had been based on sound reason: The atmosphere was looking more and more complicated with each experiment and research paper, and there were chemical reactions and air-mixing considerations coming to light that were never contemplated by the clean and certain chemical processes of the original Rowland-Molina theory. Some of the newly discovered reactions involved gases such as methane, carbon dioxide (CO_2), and nitrogen dioxide (NO_2), and under certain conditions they might greatly offset the effects of additional chlorine in the upper and lower stratosphere. By 1984, the picture of what actually goes on among the molecules of the lower Stratospheric soup (versus what computer models and experiments *predicted*) was so muddy and uncertain that no one could maintain with unchallenged authority that the net effect of CFC–borne chlorine was a provable disaster. Certainly there would be *some* ozone loss, but if it was only 2 or 4 percent, what real difference would it make? The thickness of the Earth's ozone layer varied naturally over time, and who could say the destructive effects of chlorine would force the ozone quantities outside of those natural cycles. Perhaps ozone can vary as much as 5 percent without major damage to people, plants, and animals below from increased UV-B irradiation.[4]

In the United Kingdom, James Lovelock had become similarly convinced that the problem was not terribly serious. As the years had passed, he had watched as others seemed to scale down the projected effects of the Rowland-Molina reactions more and more.[5] The implication that the warnings of major CFC–caused damage to the ozone layer might not be as serious as originally thought seemed to support his Gaia theory—the concept that the Earth is a living system that can handle anything we humans can throw in the air, including man-made CFC's. Until Halley Bay and Joe Farman's paper, there seemed no ready contradiction of that concept.

In Susan Solomon's case, it wasn't that she didn't believe Sherry Rowland or Mario Molina anymore—the concept of using the stratosphere as a CFC/chlorine dumping ground was by definition wrong in her view—but with all the uncertainties and growing complexities of the Earth's gaseous envelope and all that goes on inside it, continuing the fight was worse than tilting at a windmill. In fact, it amounted to tilting at the wind.

Sometimes, of course, it seemed that everyone in atmospheric science were doing exactly that. Nature gives up its secrets reluctantly, and complexities of figuring out the effects of anthropogenic (human-caused) pollution was no exception to the rule. There were, it seemed, no simple answers in the atmosphere.[6] Just as there were no simple methods of getting policymakers to act, and the public to listen, without a bona fide disaster, or something close to it.

High on a hill overlooking Boulder, Colorado, Ralph Cicerone—who had helped fight the same battle as Rowland and Molina—greeted the lack of substantive reaction to Joe Farman's paper as regrettable, but predictable. Then the director of the Atmospheric Chemistry Division at NCAR (the National Center for Atmospheric Research), he had learned of the paper before it hit the pages of *Nature,* and knew what was ahead for the British scientist—a steep road to acceptance of either figures or theory.

Susan Solomon had worked on her doctorate at NCAR, and if she craned her neck and almost put her nose to her office window, she could see it a few miles distant—an impressive complex of buildings looking like a modern architect's vision of a contemporary mountaintop castle—sitting at the foot of the Flatirons, the almost vertically tilted rock slabs that mark the foothills of the Rocky Mountains bordering Boulder's west flank. Susan and many of the NCAR scientists, including Cicerone, discussed the Farman paper during the summer, and how hard it was going to be for Farman to get people to take his findings on faith alone. What everyone was having a hard time getting past was the inescapable fact that the NASA satellites that watched ozone levels all over the globe had returned no unusual figures from Antarctica.

Or had they?

Deep within the NASA Goddard Space Flight Center at Greenbelt, Maryland, Dr. Donald Heath had read the Farman paper with deep distress, as had the rest of his research group. The radioed ozone data from NASA's Nimbus-7 satellite came to them, yet they had detected no ozone "hole." The group scrambled back to the voluminous reams of computer printouts to look for clues, including a rapid analysis of the program that aimed the satellite and interpreted the telemetry—and it was there they made a chilling discovery: In order to focus on the expected and normal ozone ranges, they had programmed the computers that decoded and interpreted the satellite signals to ignore and discard any reading below 180 Dobson Units. (There had been no "normal" readings as low as even 200 at that time.) But Joe Farman's Halley Bay had been recording readings around 140 Dobson Units. Both the Nimbus-7 TOMS instrument and the SBUV instrument *had* reported the dangerous loss of ozone on the planet, but no one was listening.[7]

Heath and his fellow scientists were frantic to reprocess the data, as well as go back to previous years to try and get a handle on what was really happening. Vast supplies of printouts were retrieved, and slowly they began to reconstruct what the eye of Nimbus-7 had been seeing all along as it flew silently past the Antarctic region some six hundred miles in space.

Suddenly, in front of them, like scales dropping from their electronic eyes, the Antarctic ozone hole emerged, larger than the continental United States in breadth, far deeper, and vastly more extensive than anyone had thought. And when they assembled the data from the previous years, they found themselves watching the birth pangs of a huge maw over the South Pole, a grinning void of ozone-depleted airspace they had failed utterly to detect.[8]

The British Antarctic Survey had been absolutely right. There was no question now, except the one that began hitting Joe Farman in the face within weeks: "Why did you people *wait* so long? Why wasn't this data released two years ago?"

Joe Farman handled the broadsides with stoicism and good humor, though the questions stung. It would be far easier to field the criticism over the timing of his paper than to deal with what he knew would be coming, attacks from an angry collection of British chemical companies who wanted to know why he had mentioned chlorofluorocarbons in conjunction with missing ozone. Packing even more effective control over science than their U.S. counterparts, the industrial anger translated into political force, and once more the British Antarctic Survey found itself denied new funds for expanded monitoring. Hell hath no fury, it seemed, like a profitable industry scorned.

Okay, there's a hole in the ozone. How'd it get there?

The question was simple; the answer was obviously going to be very complicated.

With the satellite data confirmation of the Halley Bay reports, scores of researchers turned their minds to the task of constructing a theory that would explain the *suddenness* of the hole's appearance. By August, NASA's Don Heath had put together a multiyear series of color slides of the hole, showing them first to a scientific conference in Prague and then Salzburg, the delegates watching in awe as the comparison of October 3, 1983, with October 3 of the previous years showed the hole as a brand-new feature, a yawning void in four colors that wasn't there before.

The thing that put a chill down everyone's spine was the speed of the reaction that had to be occurring. The common wisdom had always held that ozone losses would be steady and linear. This was not only nonlinear, the curve of increasing ozone destruction as soon as the sun came up and shone on Antarctica each September simply climbed off the charts.

My God! If ozone loss could go nonlinear there, could it happen over the mid-latitudes? Could it happen over Boulder? Was it already happening over the Arctic, where (unlike Antarctica) human populations exist below?

Or is the process absolutely unique to the temperatures, the atmospheric content, or the peculiar vortex effect of the South Pole? The question raised one of those rare challenges that can thrill scientists in a scary sort of way, because it was anything but esoteric. The answer to this one could impact the entire world.

Susan Solomon, for one, fell instantly into a dilemma. For several years she and Dr. Rolando Garcia of NCAR had worked together in trying to link the chemistry and dynamics of the atmosphere through methods such as constructing one of the first two-dimensional computer models of the atmosphere. That put her right into the heart of *dynamics* (the study of air mass movement and its effects). But Dr. Susan Solomon is an atmospheric *chemist,* so by 1985 she had become known in atmospheric circles as "the chemist who's trying to do dynamics." With that sort of reputation, it was more or less assumed that she at first would be looking for a dynamics-related cause to explain the ozone hole. Yet, the more she looked at the Farman data and at other ideas and papers (including work that Sherry Rowland and Don Wuebbles had done on the presence or absence of two molecules that could tie up chlorine, HCl and $ClONO_2$), the more convinced she was that chlorine-based chemical reactions were destroying the ozone over Antarctica, not the movement or mixing of air, even the air within the massive, seasonal polar vortex. Meanwhile in other labs, including that of Mike McElroy of Harvard, dynamics were being pursued as well as chemistry in the search for the cause.

One thought was nearly universal: Why had none of the models predicted such a rapid decrease? For a decade the atmospheric science community had crawled all over the Rowland-Molina theory looking for ways to modify, prove, disprove, or go beyond it, but there had been no widespread consideration of the horrific possibility that an ozone-eater could go nonlinear.[9]

As a result, the reaction among atmospheric scientists by the fall of 1985 was approaching the nonlinear as well, expressed in great puzzlement alternating with consternation and underlaid by frenetic research activity and deep thought.

And there was a very troubling, destabilizing point that had harpooned the previous half decade of apathy: The fact that ozone destruction could go nonlinear *anywhere* blew away the fanciful guarantees that a major ozone loss would first be heralded by an early-warning phase (during which, presumably, all the previous skeptics would suddenly join together in demanding an immediate CFC shutdown). How could anyone rely on an early warning of ozone layer damage if things could deteriorate that fast? The question flew in the face of anyone who had counseled a go-slow approach to CFC regulations (or CFC bans). The chemical and CFC industry (which had become involved in funding some of the Dobson monitoring stations), the government, and the bulk of atmospheric scientists had made a key assumption not only that there would be an early warning, but that someone would "see" it in time. That, in turn, assumed that political Washington, the Reagan administration, and the CFC industry might actually believe it. And, of course, the entire concept depended on conveniently ignoring the Rowland-Molina-Cicerone-Stolarski cautions that even a complete halt in the production of CFC's would have no real effect on the Stratosphere for forty to fifty years.

Being wrong carries its own penalties. Being stupid in the process can be very embarrassing, and though seldom publicly admitted, there were those who in looking at Halley Bay and supporting satellite data, and in remembering the warnings and the corresponding demands for proof along with the previous eight years of benign apathy, couldn't help feeling stupid. The captain of the *Titanic* must have had similar thoughts.

The first casualty was confidence in the world's basic ability to know what was happening with its vital ozone layer. The fact that Joe Farman's Survey had been the only part of the Dobson ozone-watching network around the world that had seen the massive Antarctic hole developing (and the realization that even then it took three years for him to trust his own data) didn't say much for worldwide ozone monitoring. As the media began joining the fray in late 1985, it was very apparent that the whole thing was deeply disturbing to everyone involved. It was as if mankind had suddenly reread its contract with nature under the section entitled "Ozone Layer Protection of Earth-bound Life," and been shocked to find a small clause in fine print stating: "Subject to change without notice."[10]

Of course, Sherry Rowland had suspected it all along. Basic chemical processes may be complicated and disturbed by other things, but, they are still basic chemical processes, and the world's Stratosphere was being loaded with unnatural (anthropogenic) chlorine. Rowland's quiet and steady ways and his soft-spoken professionalism had masked the frustration that bordered on cynicism at times. But he knew instinctively that despite the worldwide apathy of the pre–Halley Bay period, the ozone depletion problem had not gone away, and overall worldwide losses had probably occurred already without detection. After all, no one really knew how to monitor the total ozone content of the atmosphere with perfect accuracy.

The world had only about forty continuously operating Dobson stations watching the ozone layer by 1980, and it was a bit of an apples and oranges mixture. The instruments were owned and operated by a wide variety of nations and institutions, and maintained in different ways at different intervals. The so-called Dobson Unit—the unit of measurement used by the often-cantankerous instruments—was basically the same from station to station, but there was very little international coordination, and it was not only difficult but scientifically risky business to try to compare the data from dissimilar Dobson monitoring networks.

Yet that's precisely what had been happening, with the enthusiastic blessing of the chemical and CFC industry. In the early eighties several worldwide ozone trend analyses were produced by various groups by comparing the Dobson data from around the world, a task accomplished principally by lumping the data for all seasons of the year together—as well as for all latitudes monitored—and averaging everything.[11] Not surprisingly, then, the trend analyses had all shown "no statistically significant change in natural behaviors."

Sherry Rowland could be expected not to trust such results, and he lived up to that expectation, especially when he discovered that one of the oldest and most reliable stations—in Arosa, Switzerland—had recorded an 8 percent reduction in "total column ozone" (what the station can see overhead averaged for all altitudes) in 1983 that had not yet been included anywhere in the latest ozone trend analysis. In early 1986—after Joe Farman's paper had come to his attention—Rowland dispatched one of his graduate students, doctoral candidate Neil Harris, to dig into the real figures from Arosa, and within a few months they had discovered similar supporting readings from several other northern-latitude stations in Maine and North Dakota as well as West Germany and Canada.[12] Each of those places had shown a corresponding drop in 1983, but the drops had disappeared in the corrections and averaging of the statistical trend analysis for 1983, which Rowland was convinced had not been subjected to sufficient scientific peer review in the first place. When they looked even closer, they began to find a discernible pattern of winter losses and summer gains in ozone, as well as patterns of loss and gain at different altitudes. When you averaged it all, it appeared ozone levels hadn't changed. But when you looked more carefully, ozone losses *were* occurring—and increasing—though not on any nonlinear basis.

There was an additional plot twist to the puzzle of 1983's readings that would deeply worry Susan Solomon and others several years later. The Mexican volcano El Chichón had erupted in April 1982, injecting thousands of tons of material and acids into the Troposphere *and* Stratosphere; and, according to the data developed by Rowland and Neil Harris, within months you could find ozone losses over the northern latitudes. Could there be a connection? In the aftermath of Joe Farman's bombshell paper, the question was suppressed for a while, but not forgotten.

It had long been assumed that really significant chemical reactions occur only in uniform mixtures of molecules. (In other words, if you have 21 percent oxygen in a uniform mixture in a given block of airspace, you can safely assume that *everywhere* within that airspace—no matter which slice or section you sample—you'll find 21 percent of the molecules are oxygen.) Chemical reactions among uniform substances proceed with a certain degree of easy predictability, but non-uniform substances are by nature more *un*predictable, with reaction speeds that may be much greater.

The sudden presence of El Chichón's volcanic remains in the Stratosphere and a suspiciously corresponding drop in ozone readings the next year suggested to both Sherry Rowland and Mario Molina (working separately) that maybe something other than the traditionally modeled homogeneous reaction was taking place. Perhaps—just perhaps—some strange reactions were taking place on the surface faces of tiny particles from El Chichón—a *heterogeneous* reaction. Could that speed up ozone destruction? (Rowland, in fact, had been pursuing aspects of this type of reaction for several years as he looked for the possible effects on various atmospheric chlorine sinks.)[13]

Rowland and Molina both began doing experiments to find out whether the idea had merit. They had talked about the idea years before, and done little with it. Now they began to find that reactions they were familiar with in a gaseous state could be greatly speeded up on the surfaces of laboratory containers (made of glass and Teflon). Mario Molina published a paper on his findings in early 1986 while Sherry Rowland and Susan Solomon both began to wonder if heterogeneous reactions could have something to do with the Antarctic ozone hole.

There were other clues as well keeping Susan's attention riveted on the subject. A team of scientists in New Zealand had published a paper in 1984 showing curiously low levels of NO_2 (nitrogen dioxide) over Antarctica, and that could be significant, she knew, because NO_2 is an important reservoir for chlorine, tying it up and preventing it from eating ozone. *If* something was happening to reduce the NO_2 in the atmosphere down there, it could dovetail with all the other curious bits and pieces to form a solution.

By early 1986 Susan and Rolando Garcia had been joined by Sherry Rowland and Don Wuebbles of Lawrence Livermore Laboratory in constructing what seemed like the most plausible assumption and explanation of what was causing the gigantic hole in the ozone layer centered over the South Pole each September through November (a proposal that would take the form of a paper subsequently published in June 1986).[14] The images of polar stratospheric clouds and the idea of heterogeneous reactions had coupled with the reality of extremely cold temperatures in the Antarctic night and the potential denitrification of the air to produce the idea that the ozone hole was indeed caused by chlorine chemistry (resulting from CFC–borne chlorine), but the conditions were very strange and unique to Antarctica.

The PSC's—the beautiful polar stratospheric clouds—were the key, they figured. The tiny ice crystal surfaces were like little crucibles that let reactions involving NO_2 accelerate drastically, leaving little NO_2 around to act as the keepers of chlorine.[15]

But how?

The theory, in fact, would not be well received, and even one of Mario Molina's colleagues would later write Susan Solomon to explain that the process they were postulating was extremely improbable.

In his lab at the Jet Propulsion Laboratory, however, Mario Molina had already begun turning his attention to a series of reactions he and Sherry Rowland had been talking about for many years, reactions on heterogeneous surfaces that might explain what was going on in the purple darkness over the southern ice cap.[16]

Using the same quiet brilliance and thoroughness that had buttressed and refined the original Rowland-Molina theory since 1974, Mario began staging a sophisticated series of reactions designed to find out how chlorine could be freed from the stable chlorine reservoir known as hydrochloric acid (HCl). Tiny droplets of hydrochloric acid floating in the stratosphere amidst the PSC's "absorbed" a considerable amount of free chlorine, locking it up and keeping it from doing damage to ozone. But if the hydrochloric acid was somehow being converted in the frigid night into molecules that could be broken down by sunlight in the Antarctic dawn, then the effectiveness of the hydrochloric sink would be lost. That seemed to be happening, but no one had figured out how.

Molina began forcing chlorine compounds through a small glass tube past a film of ice, and found that hydrochloric acid and chlorine nitrate (another molecule existing in the stratosphere) could react on the surfaces of the very type of ice crystals found in PSC's, leaving the chlorine ready for release at the first rays of sunlight. And more important, Mario and his team discovered that contrary to existing belief, that chain reaction occurred very rapidly.[17] Those new facts constituted a major breakthrough in understanding the Antarctic ozone hole's possible chemistry, but it wasn't enough. There was still a gap in the explanation, a missing link in the catalytic chain, and through more lab work and a brilliant intuitive leap, Mario Molina found the key.

The reactions of chlorine compounds on the tiny ice crystal surfaces of the PSC's set the stage all right, but there was an odd doublet of chlorine monoxide known as a ClO "dimer" (ClOOCl) that gave the process of ozone eating the speed it would need to destroy ozone at an exponential rate. Mario not only proposed the dimer as a key element of the ozone destruction chain reaction, he demonstrated it in the lab, showing that the formation and asymmetric decomposition of ClOOCl was the prime step that enabled ClO to destroy ozone in the unique conditions of Polar Stratospheric Clouds and the Antarctic Stratospheric soup.[18]

And the end result of such exotic chemical reactions? A sky full of free chlorine with no remaining enemies. It was like stacking dry tinder in preparation for a bonfire. Once the sunlight returned to the region, the chlorine would be free to start chewing energetically through the ozone, reducing it to stable oxygen. Of course, it was all theory until someone could bring back the actual evidence from the perturbed Stratosphere over Antarctica.

In the spring of 1986, however, Molina's findings were still months away, and as Susan Solomon began slaving over a hot word processor in her Boulder office to finish her paper with Rowland, Garcia, and Wuebbles, another professional friend was heading to town to further change her life.

The scraggly beard and wild, undisciplined black hair were familiar features to his colleagues, but outsiders who encountered Dr. Robert Watson of NASA before hearing him speak were sometimes taken aback—as if they'd finally met a "mad scientist."

The voice changed things, somehow, the rapid-fire energy and enthusiasm of the man wrapped in his British accent that tends to steal all attention away from physical appearance and rivet it on what he has to say. Watson was the head of the NASA's Stratospheric Research programs, and a highly respected scientist.

Now Watson was back in Boulder for a March 1986 conference on atmospheric ozone organized before Joe Farman's revelations about the readings at Halley Bay had become the number one passionate topic on Susan Solomon's mind. The conference had been called to discuss the need for worldwide monitoring of other molecular species than just ozone (nitrogen dioxide and hydrogen chloride were also factors that few people were tracking), but there had been no time set aside to discuss the Antarctic situation.

Susan Solomon took the opportunity to lobby Bob Watson for a special half-day session on Farman's findings, and Watson agreed. They needed his catalytic help in figuring out what to do—where to go from there. And Susan in particular wanted to know who was going to Antarctica and when—they were in dire need of data to provide the grist for the scientific mill.

As the meeting progressed in one of the conference rooms in the main NOAA administration building at Boulder, Watson suddenly turned to Art Schmeltekopf of the NOAA Aeronomy lab, one of Susan's colleagues, and asked, "Can your instrument do OClO [chlorine dioxide]?"[19]

Susan overheard the question and the speculation that the answer might be affirmative, and it thrilled her. Chlorine! That's where all this was bound to go anyway. But measuring chlorine compounds would be critical in validating any theory.

During lunch Watson and Schmeltekopf disappeared into the library to look up the spectrum of the OClO molecule to make sure his instrument could track it.

It could. They had the instrument. Now they needed the means of getting it to Antarctica, not to mention a team to go with it (as well as other measuring devices and the scientists to operate them, of course).

Bob Watson had already been polling people at the main meeting, trying to find out what, if anything, could be done the coming summer (the southern winter of 1986). "Not much this year," had been the standard response, yet Watson refused to give up. He knew there was at least one way to get to "the ice" in winter, though it could be brutal.

At a meeting at NASA Goddard Spaceflight Center in Maryland in April, Susan and Sherry Rowland had insisted to Bob Watson that any expedition had to go in August. October would be too late. Observations were needed while the ozone was being lost in early September, not after it was already gone. The difficulty, of course, was that August was the dead of the Austral winter.

Watson was well aware of the fact that winter in Antarctica is a lonely ordeal, a trial-by-ice few scientists ever experience. There is no regular air service, no usable runway from May through September, and no way out except in a dire medical emergency during the dead of winter. America's main research outpost, McMurdo Sound, receives an airdrop of fresh food and mail in June, but other than constant satellite communications links, it might as well be in space somewhere, alone and unreachable—literally at the end of the Earth.

In late November the annual migration of scientists return to their research projects on the ice, arriving in droves at McMurdo Sound—which gives the few military and contract personnel who've kept things running through the winter the chance to rotate home again. To get everyone in, however, requires prepared ice runways, and much of the equipment to get those ready comes in by air during September, still in the dead of winter, on board ski-equipped turboprop C-130's operated by the Navy for the National Science Foundation—lumbering and uncomfortable airlifters that are flown out of Christchurch, New Zealand, on a nine-hour trek to the ice. These winter-fly trips are few and far between, but Watson calculated that the National Science Foundation and his own NASA might be sweet-talked into letting them hitch a ride in on a winter-fly mission during August.

That would be the easy part. Assembling a scientific team with workable instruments would be the hard part.

Watson knew there were certain people who should go to Antarctica, whenever an expedition was mounted, and one of them was Dave Hofmann of the University of Wyoming at Laramie, who was already a well-traveled hand at Antarctic research. Hofmann had developed balloon-borne instruments that could measure some of the same molecular compounds they were searching for in the polar stratosphere, and he was preparing to go back to McMurdo anyway in November. Moving him up three months was within the realm of possibility. Barney Farmer of the Jet Propulsion Lab in Pasadena was another important member, because of his multimillion-dollar space-shuttle-based instrument that could do some unique measurements from the ground in a slightly different way. Farmer would not be anxious to have his expensive spectrometer hauled to Antarctica, but it was worth a try to ask him along. Whether either the scientist or his equipment could do it was another question.

There was a team from the State University of New York at Stony Brook who had built a microwave instrument that could measure ClO at all altitudes, and they too were a prime candidate.[20]

The process of getting instruments ready in time—and having some hope they'd work in the incredibly cold temperatures—would be daunting. It would make no sense to send a team south if they had no hope of making *some* measurements (even though the standard wisdom was that people never got good data on their first trip to the ice).

As Watson probed his assembled colleagues in Boulder, several said they wouldn't be available for a wide variety of reasons, and others had colleagues to check with and funding requirements to consider. Even Art Schmeltekopf demurred (because of prior commitments to a tropical experiment that year), and he was crucial to the mission if his instrument (a diode array spectrometer) was to be of any use. *Someone* would have to go along from the NOAA Aeronomy Lab to operate Schmeltekopf's machine if he couldn't do it.

Just when the group fell silent, leaving Bob Watson without a volunteer, a familiar voice—tinged with a bit of resignation—wafted through the room.

"Okay," the voice said, "I guess *I'll* go."

Susan Solomon was more startled than anyone else in the room at the origin of the voice.

It was hers.

CHAPTER

4

In the Belly of the Beast

McMurdo Sound, Antarctica, August 1986

The blast of utterly frigid Antarctic air was truly amazing—like opening the deep freeze in a biology lab. It had enveloped Susan Solomon suddenly like a slap in the face with cutting sharpness and depth, a shocking absence of warmth—as if nature was giving her a personal lesson in what those little digits preceded by a minus sign *really* meant in the real world.

At least the real world at the south end of the planet.

She hadn't been prepared for −45°F when the rear cargo ramp and doors of the Lockheed C-130 opened, destroying the semi-comfortable cocoon they had occupied for eight hours. The four Allison turboprop engines were slowly winding down now, their internal fires extinguished, their huge four-bladed paddlelike propellers in slow motion and approaching a stop, and Susan began coming to grips with exactly where on Earth she was.

The Navy crew members of the C-130 were wasting no time in starting their down-load "on the ice," removing the heavy cargo chains and the locking devices on the large square cargo pallets that spanned the aircraft floor, each pallet piled high with boxes and bags covered by heavy straps and netting. As soon as each pallet was unlocked, the loadmasters would push it to the back of the airplane, the side rails guiding it along the roller-equipped cargo floor to the near-frozen metallic arms of a waiting forklift.

So this is Antarctica! The thought had to be acknowledged, since the human response she was feeling was very primal. Susan looked around at the landscape of total whiteness, aware of the tectonics of the place, aware that she was standing on the very edge of the Ross Ice Shelf, looking across the bay to the west at land and rock and the Prince Albert Mountains of the Scott Coast. Unlike the Arctic, there was an actual continent with real rock here, much of it beneath hundreds of thousands of years of accumulated ice—though some coastal portions of Antarctica, such as the one they had used as a runway, were actually ice shelves of impressive thickness with nothing but water below, similar to the ice cap over the North Pole. It had been seventy-five years since Roald Amundsen beat Robert F. Scott to the South Pole, Scott following a month later with far heavier equipment that eventually doomed his expedition, his men, and him to an agonizing death. Now, where Robert Scott had begun his trek, there stood a regular scientific outpost—a large base, really—with scores of drab, conventional buildings, a few Quonset huts, and enough dormitory-style facilities to house and feed and provide working space for over a thousand people. McMurdo, Susan knew, had communications facilities, lab facilities, cafeterias, and all the necessities.

But it was still Antarctica, and there was a surreal element to all of this. Geologists and geophysicists were used to ending up in strange, hostile places in search of their truths, but chemists used to modeling atmospheres and happily attending scientific meetings to further the intellectual ferment of atmospheric chemistry—scientists like her, in other words—seldom had to contend with anything more hostile than poor hotel restaurants and the ubiquitous, joyless hassles of deregulated air travel.

Susan glanced at the loaded cargo pallets with some apprehension, the urgency of the task ahead of her competing with the desire to get inside somewhere and away from the freezing cold. There were some very delicate instruments on those pallets that might not survive subzero temperatures. The temperature-sensitive boxes were all painted black, and they littered the snow as the unpacking continued. Each one would have to be brought inside rapidly, and that would probably take the rest of the day, way into the evening hours. Evening, of course, was a relative term. There had been only dim twilight on arrival—now it was dark again.

Susan was in charge of the first National Ozone Expedition (NOZE for short), but she and the other twelve scientists all worked as colleagues, and there was no need for militaristic orders or directives. Nor, apparently, was there any worry about getting people in motion. In the deep freeze that surrounded them, there was a rather impressive, built-in incentive to keep moving to keep warm.

The one hundred or so winter residents of McMurdo—mostly civilian employees of ITT Antarctic Services (the contractor company that kept the place going) and some Navy personnel (all enthusiastic volunteers)—had been waiting for them when the big turboprop taxied to a halt. As soon as they got close enough, the McMurdo crew noticed that the NOZE-1 team leader was an attractive female (after months of not seeing a woman that was hardly surprising). Once she left the C-130, Susan's dark hair and large, brown eyes were almost invisible beneath the furry hood of the Arctic parka issued to her in Christchurch, New Zealand, but no matter. They watched her with great interest anyway.

All the residual heat of her body seemed to have been sucked away by the painful cold. Susan hated cold weather, or so she had always told herself. She was a scuba diver who loved warm weather, warm water, balmy breezes, and bright sunlight (properly filtered of UV-B radiation, of course). So what in heaven's name was she doing in McMurdo? The question had popped up several times over the long hours as she sat amidst the deafening drone of the C-130's turboprop engines, her ears and head in the grip of the massive sounds and vibration in an aircraft cabin never intended for passenger comfort—riding sideways on the stretched red canvas that passed for sidewall seats.

Talking in the C-130 had been a near impossibility, so they had spent the time sleeping and reading. Now fatigue was a genuine factor, especially after a midnight wake-up call in the hotel near Christchurch airport for the second attempt at the trip (on the first they had had to turn back when one of the retractable skis on the lead C-130 malfunctioned). Her band of thirteen had inspected their personal possessions and their research gear (nearly fifteen thousand pounds of it), and climbed aboard again in predawn darkness, bleary-eyed but excited.

The flights south were timed to arrive at McMurdo during the brief period of twilight, they had been told, just in case they had to make a forced landing. It was one of the finer points of flying south—along with the fact that a second C-130 would be going along with them to serve as a rescue ship in case the first plane had a problem past the PSR, or Point of Safe Return. Flying to McMurdo was not an especially dangerous act, but most pilots considered it an *unnatural* act, and one that demanded extraordinary attention to safety precautions—even more so in the winter. After all, McMurdo was 2,100 miles from Christchurch (a 4,200-mile round trip), which is just about the range limit for a C-130. Once they had passed the PSR (usually called an ETP, or equal time point), they were committed to land *somewhere* on the ice, with or without a runway or visibility. It was considered a far-out possibility, but whiteouts (total loss of visibility in blowing snow and ice) had occurred before at McMurdo without warning, and without enough fuel to go anywhere else, the four-engine C-130's would have to land blind, feeling their way down to the ice somewhere safely distant from McMurdo, descending at no more than a fifty-foot-per-minute rate until the skis touched the surface, while hoping there was nothing but flat ice in their path. It had happened before—an airplane forced to land blind and straight ahead on the ice when the pilot no longer had other options—and the account of that ordeal in the book *Southlight*, by Michael Parfait, had been scaring all of them for the past few days.

The runway at McMurdo was nothing but ice and snow, of course, and even though they had heard that the ITT contractors at McMurdo would have already spent the better part of a day grading it, winter landings on the ice were physically rough. Normally there were six winter fly-in's courtesy of the Navy (two additional ones had been added to accommodate the NOZE-1 team), and the last series of the season would bring personnel in to get the runway in better shape for the faster, larger (and comparatively more comfortable) C-141 Starlifters flown by the Air Force's Military Airlift Command (MAC)—the airlifters used to ferry the scientific community back and forth during the Antarctic summer.

On this trip, however, the cargo was scientific, and the mission was of potentially vital importance to the Earth as a whole.

The fact that the expedition was coming had been generating excitement for some time at McMurdo. They were well aware from news reports of the gigantic ozone hole above them, and the select team that would try to find out what was happening and why it was there were going to have them as hosts—and eager hosts at that. As Susan and her cohorts had stood blinking and trying to get their bearings moments after arrival, they were greeted by an enthusiastic "can-do" attitude that would never flag over the next two months. NOZE-1 had arrived—and it had been a bit of a miracle they'd gotten this far.

Bob Watson had accepted Susan Solomon's offer on the spot during the spring meeting at Boulder, but it was several weeks later that she was asked to head the expedition, taking charge of the administrative and organizational aspects. The moment Joe Farman's data had been verified by the NASA satellite readings and the existence of the ozone hole was accepted as fact, the need for more information, readings, and observations had become acute. There was no question that some sort of expedition to Antarctica was needed, but when? Such things take tremendous amounts of time, effort, money, and coordination to piece together. You don't just buy a set of airline tickets and send your people south. There were sensitive instruments to prepare, complicated and expensive logistic arrangements to plan, and funding to secure—not to mention the need for coordination with the military, McMurdo, and the National Science Foundation (which controls the American scientific missions to the ice). And all of that was based on the not-necessarily-sound assumption that Bob Watson could wheedle permission from his own organization, NASA, to mount the last-minute expedition. How all that could possibly be done by the Austral spring of 1986 (which at that time was a mere five months away) was a question with only one answer: "If anyone can pull it off, Bob Watson can."

Watson was known for doing the impossible. In fact, with his unremitting energy and reputation for being a dynamo with limited understanding of the word "no," his ability to blend the diplomatic and administrative with the scientific was somewhere between rare and unparalleled. He seemed to thrive on organizational challenges, overwhelming bureaucratic roadblocks by force of logic and reason cloaked in diplomacy and personal charm.

"He's a bearded Sara Lee," says one friend. "Nobody doesn't like Bob Watson."

But if ever Watson had a worthy challenge, this was it. The urgency of finding out if CFC's were at fault was obviously very real, and worldwide scientific jitters over the ozone hole and what it meant demanded a dedicated and immediate research effort. Theories of what was causing the hole were already proliferating, and if more precise and reliable data couldn't be provided rapidly, the atmospheric scientists of the world would tend to migrate even further into diverse theoretical camps, hardening their positions, lessening cooperation with each other, and slowing the process of reaching a sufficient consensus to trigger international action (if the cause was indeed CFC–borne chlorine). It was exactly that type of polarization that had traditionally prompted both government and industry leaders to walk away from fractious scientific debates in disgust with a parting shot: "Give us a call when you can agree on something." There was too much at stake now to let that happen again.

And there was far too much uncertainty about what was happening to the Antarctic ozone.

Susan Solomon, Sherry Rowland, and many others suspected the ozone loss stemmed from chemical reactions involving chlorine, but there was another school of thought that unusual, changing patterns of air circulation around the South Pole (which had nothing to do with human contamination) were transporting ozone-poor air into the region and simply giving the false appearance of ozone loss by ozone destruction. (This was rapidly labeled the *dynamics* class of theories.)[1]

There was also one group of scientists who suspected that solar activity—sunspots and flares—were triggering nitrogen reactions that caused ozone destruction, and the ozone hole.

And of the three, only the chemical theory pointed a finger at man's production of CFC's. If either of the other two theories were correct, it would mean the hole was natural, and the ozone would eventually recover on its own. But if chlorine *was* the culprit, action was needed to put a cork in the bottle of growing worldwide CFC production—fast.

By May, Susan Solomon had known that *if* the Solomon-Garcia-Rowland-Wuebbles paper had things figured out correctly, and *if* they could look clearly into the Stratosphere from McMurdo with ground-based and balloon-borne instruments, and *if* they could spot concentrations of chlorine monoxide (ClO) and dioxide (OClO) and several other substances in the predicted quantities, they just *might* be able to give an initial indication of the cause—or at least narrow it down to a smaller set of theories. High concentrations of ClO and OClO, very low concentrations of nitrogen dioxide (molecules that could lock up the active chlorine in reservoirs)—and the presence of Polar Stratospheric Clouds in very low temperatures—would add up to the frozen witches' brew they theorized could cause a nonlinear ozone loss as soon as sunlight hit the mixture. It was a good theory, but they had a long way to go to validate it. And without good observational results from Antarctica itself, they hadn't a prayer.

Now, suddenly, they had a chance to make those measurements a year ahead of schedule.

Susan had been amazed at the energy the ITT people were using to help them unload, but she figured her team would be left on their own after 5 P.M. quitting time. After all, these were *civilians*. After dinner, however, the ITT and Navy people followed NOZE scientists right back out the door without a second's hesitation and kept on moving boxes until the job was done. It was nearly midnight when they got the last of them secured.

The accommodations were somewhere between a defrocked Holiday Inn and a college dorm with bathrooms down the hall: clean and livable, but nothing to write home about—even if there had been mail service from McMurdo in the winter.

Susan had half shuffled into the drab little room about midnight—cold, exhausted, a bit disoriented, and apprehensive. The depressing enclosure would be "home" for over six weeks, but that was an insignificant matter. What was bringing her down were the warnings—and the practical realization—that despite all the effort, her NOZE-1 team probably wouldn't get the data they needed on this first trip to the ice.

They had been told over and over again: "No one's experiments ever work the first time at McMurdo." In the morning they would get a look at whether the prediction extended to the instruments themselves, and whether they'd survived the trip. She worried in particular about Art Schmeltekopf's spectrometer (a diode array spectrometer), which was supposed to be her primary experiment. What if it were rendered useless? She and her colleagues could *run* the thing, but if it were hopelessly broken nine thousand miles from Boulder . . .

They were all concerned about Barney Farmer's multimillion-dollar "box." He had been brave enough to take his extremely sensitive and important apparatus and haul it to the end of the Earth, but it was anyone's guess whether it would—or had—survived the trip well enough to give them useful readings. The ITT men had treated it delicately enough, placing it in a specially prepared shed across from her dormitory, but several days of setup and testing would be necessary before they knew whether the effort had been in vain.

At least the Navy C-130 pilots had not been forced to dump everything on the ice and run. There was a procedure, they had been warned, that flight crews on the ice would have to follow when a whiteout was approaching or the weather conditions were deteriorating rapidly as they landed at McMurdo. The crew would open the rear cargo doors, unlock the pallets, drop the ramp to surface level, and just let the cargo slide off and drop onto the ice as the aircraft taxied along and prepared to leap back in the air. It was a brutal procedure certain to damage at least some of their sensitive gear.

But the day had been clear, and since the C-130's had to

refuel at McMurdo for the trip back to Christchurch, there was time to lower the pallets to the frozen surface as gently as forklifts and cold hands could manage.

Of course, everything inside those pallets had been riding a four-engine vibrator for the past eight hours, along with being shaken, jostled, jerked, and generally rattled around since leaving the states. Such treatment was unavoidable, but the specter of arriving at McMurdo with nothing operational but their notebooks haunted all thirteen of them.

And there were the overall, industrial-strength worries sure to beset a team leader as soon as the frenetic arrival activities were over and the real purpose of the mission lay dead ahead: Would the unique polar vortex be anywhere near McMurdo during the next week, for instance? (The ozone losses occurred *inside* the vortex—peering into the air *outside* without corresponding interior data wouldn't be useful.) Would they have clear weather at McMurdo? (In bad weather, only the balloon-carried instrument packages could be used and limited ground-based data acquired, and if the winds were too high, even the balloons were useless.) And would the results be sufficiently unambiguous? (NASA and NOAA, NCAR, the University of Wyoming, the National Science Foundation, the U.S. Navy, and even the Chemical Manufacturers' Association—which contributed funding—had expended a lot of time, effort, and money to get the little band of thirteen to McMurdo. What if it all turned out to be a waste of time?)

If Susan hadn't been exhausted, such questions would have kept her awake.

Tuesday, September 16, 1986, had dawned clear and frigid over McMurdo Sound—though it wasn't much of a dawn. At 7 A.M. it was bright twilight above—the hint of a twilight glow on the horizon that Susan had barely noticed as she and the others fine-tuned their equipment. Barney's Barn—the name given to the shed housing Barney Farmer's Atmospheric

Trace Molecule Spectrometer (which had survived)—was across the way from the lab building where Susan's team had set up. By the time the noon sun had come and gone, Schmeltekopf's spectrometer had been positioned below a small hole in the roof—a hole that still amazed her.

It hadn't been there before, that hole, yet they had needed some method of looking straight up while keeping Schmeltekopf's equipment in warm room temperatures. That meant cutting a hole in the roof of a government building.

"You need a hole?" one of the ITT men had asked.

"Yes," she had replied, bracing inwardly for the response she knew was coming—the response that *had* to come from a government contractor asked to alter the essence and form of government property without something approaching congressional authority.

"No problem," he said. "*Where* do you want it?" And within hours Susan's team was in business and operating the spectrometer.

Of course, that in itself was a thermal agony. There was nothing wrong with the spectrometer, but the method of getting light to it was an exercise that might have made Rube Goldberg proud. In a phrase, it was all done with mirrors.

They needed moonbeams. The sun's light as reflected off the moon's surface would pour onto the McMurdo area, having passed through the molecules of gaseous compounds overhead. If that light were directed down the newly installed pipe through the newly cut hole in the roof, a grating could break the light into a spectrum that could be carefully analyzed by Schmeltekopf's apparatus. Susan and her team could then look at the computer analysis of the different light bands (spectra) that searched for the telltale combination of light-and-dark areas (which look something like an irregular Universal Product Code marking, but which carry unique molecular signatures) of particular molecules.

But it had to start with the gathering of moonbeams, which sounded somewhat lyrical until the team realized that since there was no tracking device available and the pipe couldn't be angled from below at will, the mirror would have to be aimed and held by hand on the roof.

The roof was not heated—and this was, after all, Antarctica in winter. The temperatures enveloping those exiled to the roof (they took turns in twenty-minute shifts to reflect moonbeams toward the hole) were around $-50°C$, and winds could drive the effective chill factor much further toward the basement—all of which sparked dark and uncharitable speculation that the *real* reason Art Schmeltekopf had not accompanied his instrument was now achingly apparent.

Chlorine dioxide (OClO) and nitrogen dioxide (NO_2) were the targets they were searching for. OClO would indicate that ozone-eating activities related to chlorine were definitely afoot, which was Susan's best theory. Finding NO_2 in high quantities, however, would tend to indicate that the ozone loss might be attributed to natural processes caused by solar cycles.

Susan's team was focused on getting the data. No one spent time worrying that one scientist's theory might be discredited and another's reinforced, and no one seemed concerned that all the time and effort and expenditure might culminate in the embarrassing conclusion that the ozone hole was nothing more than a natural phenomenon that would eventually repair itself. That worry, however, had indeed crossed the minds of at least a few of those who had remained behind. Such a finding would be good news for Earth, of course, but it could tend to make the atmospheric scientific community appear to be semi-hysterical and given to gross overreaction.

But the first order of concern for Susan Solomon and the members of NOZE was the task of getting good data, and by the morning of September 19, it was surprisingly apparent that they were doing just that. The previous three nights had brought clear skies, a moon at just the right angle, reasonably low winds, and (they would later discover) a polar vortex that had obligingly moved over McMurdo and stayed there. Susan and her team had actually been staring up into the belly of the beast, and their readings were already beginning to implicate chlorine.

Barney's barn had similar success, as did Dave Hofmann's team—who had been very worried that the extreme cold would ruin his gossamer-thin plastic balloons and send their large instrument packages crashing into the McMurdo complex or onto the trackless ice. The balloons survived, however, and the data sets looked very promising right from the first.

The Stony Brook team from the State University of New York—equipped with a microwave device—had also reported good data, and as Susan looked around the room during their morning "how-goes-it?" meeting, the fact that they were succeeding on the first trip to the ice—and the realization of how rare that was—struck home.[2]

Of course, she had always been blessed with a bit more luck than she could explain. It was the "Susan Solomon Rabbit's Foot Effect" according to several co-workers. Take Susan along, they said, and everything just seems to work out.

Usually.

By mid-October as the first seasonal C-141 flights began arriving at McMurdo carrying the thundering herd of over a thousand summer resident scientists, the trends in the data were leaning the way the chlorofluorocarbon theories had predicted—significant levels of OClO, very low levels of NO_2 (in fact some of the lowest ever recorded in Earth's atmosphere), and enough evidence to strongly discount either the pure dynamic theory or the solar theory. It *was* apparently a chemical process, and Susan saw no reason not to say so at the appropriate moment.

Determining the definition of "appropriate" would, however, be a significant problem.

There was serious public interest in the scary concept of an ozone hole. The media had been growing increasingly aware of the problem, and the hunger for information from the

NOZE-1 team was substantial. In mid-October that hunger became too much for one newsman back in the United States long on curiosity and audacity—and short on ethics.

"Susan, you have a personal emergency phone call."

The words froze her in her tracks. She had been almost totally out of touch with her mother and father and brother since leaving in mid-August, except for two rounds of letters. The National Science Foundation guarded the telephone number to the satellite connection with McMurdo very carefully, and permitted calls only from family members when a personal emergency existed. Something was very wrong.

The call would be replaced in two hours, she was told. There was no way to replace the call before then. For two hours she wondered and worried. Thoughts of sickness, death, injury, or a thousand other scenarios flitted around the corners of her pragmatic mind. Whatever it was, she would have to handle it somehow.

Finally she called the main base switchboard and discovered that the caller had left a name. She expected the last name of Solomon.

What she heard was the last name of a newspaper reporter.

In utter disgust she told the operator to tell the man she wouldn't accept the call.[3]

The incident was infuriating. Susan spent the rest of the day working to control her anger, mindful of the fact that cumulative fatigue coupled with the phone call simply reinforced the loneliness and isolation of McMurdo.

The end of the expedition was drawing closer, and it was with some relief that all thirteen of them gathered around the telephone on October 20 to report their preliminary findings to a news conference thousands of miles away, before beginning the process of dismantling the equipment and flying home.

On the other end, in a meeting room within the impressive headquarters building of the National Science Foundation in

Washington, D.C., a speakerphone relayed Susan's voice to a gathering of newsmen. The conference had been arranged in advance as an end-of-expedition update. They were merely going to report their preliminary findings, knowing that the data they would bring back from McMurdo would take months to properly digest and interpret. The conference was not supposed to provide *the* definitive solution to any part of the ozone hole problem.

Nor was it supposed to start an internecine war.

Sherry Rowland among others had warned that there would be great pressure on the NOZE-1 scientists to come back with a definitive "verdict" on the central question: Are CFC's responsible for the hole? Yet Rowland knew better than most how difficult it would be to convince anyone in the scientific community—let alone the industrial community—with first-time data from a single expedition to one spot of the vast Antarctic continent. The billion-dollar CFC industry was hardly going to throw in the towel without beady-eyed, skeptical scrutiny of what would have to be overwhelming evidence.

Susan Solomon had her hardworking compatriots around her. She had nearly two months of exhilarating, freezing, backbreaking, and nerve-racking work behind her. All of them had scrutinized the data with the greatest care and honesty, she knew, since they were acutely aware that the eyes of the world were on them. As professional scientists, they could not afford to jump to a single conclusion.

But she also knew they had the obligation to identify spades as being spades—and chemical causes as being chemical causes—which is precisely what she did.

"We suspect a chemical process is fundamentally responsible for the formation of the hole," she said.

That meant chlorine and CFC's to the scientists at McMurdo, but to the newsmen at the NSF building back in Washington who knew there were three theories—a "chemical" theory, a "dynamics" theory, and a "solar" theory—the NOZE-1 team had just elected the "chemical" theory. That meant a contest had been lost by those with the defeated theories, and like good newsmen hampered by limited understanding of the scientific process, they trooped off to talk to the losers—who were in no mood to concede defeat.

It was an oft-repeated scenario. Journalists are trained to think bilaterally, not multilaterally. There are two sides to every issue—not three or four or more—and to be scrupulously fair, when you report one side, you must dutifully report the other at the same time. If a Republican senator blasts a particular bill, for instance, a "good journalist" will call a Democrat for a rebuttal position. If the Democrat also dislikes the bill, the poor journalist has to keep calling until he finds someone who supports it and will speak well of it—even if that someone is not very credible. Both sides deserve representation, no matter what it takes.

In journalism, an interview with God would create an obligation to call the devil for a balancing viewpoint.

But in science, there are quite often a multiplicity of "sides," and sometimes there are merely postulations, ideas, preliminary positions, and embryonic theories, none of which are diametrically opposed to any other. To a past dean of science journalism such as Walter Sullivan of *The New York Times,* this was a well-understood dilemma, and seldom would a reporter of such caliber ever make the mistake of artificially creating a battle line that didn't exist in trying to reduce a complicated scientific problem to a simple two-sided controversy. The majority of the media, however—broadcast and print—have little understanding or training in the peculiarities of reporting the scientific process. The inside catfights and rows seem more the stuff of combat than some sort of seminal, interactive process, and the result is too often the artificial polarization *in the reporter's mind* of what is, in essence, the noisy process of the scientific method.

Of course, combative scientists who have an equally abysmal understanding of the media and their needs sometimes incite the process. It's depressingly familiar to see an upset researcher (nursing an endangered thesis) flip a disparaging offhand remark at an eager reporter—like a chunk of tenderloin casually tossed to a pack of hungry dogs—then profess amazement when he gets bitten in the resulting melee. It can be very embarrassing when such remarks end up in the national media. Even scientists must learn when to say, "This is off the record."

The dynamicists were not amused, to say the least. Dr. Mark Schoeberl, for one, had been hard at work during the summer trying to fit temperature differences and air mass movements into a model that would explain the ozone loss as a function of "transport" (another word for "dynamics")—a natural function, in other words, owing no allegiance to chlorine. He was well aware that his postulated solution might cut across other lines and include some elements of dynamics and some related to chemical processes (and perhaps something exotic no one had thought of). Schoeberl and several others had made progress by late August with new nonchemical explanations, but the NOZE-1 team—on the ice and out of touch—had no knowledge of their latest ideas. If Susan Solomon's expedition had quietly returned and quietly published the very same preliminary findings enunciated by phone from McMurdo (an impossibility given the media interest), the level of response from those like Schoeberl who disagreed with a chemical solution might conceivably have been different. When not challenged in public, careful scientists seldom feel compelled to defend themselves in public.[4]

But suddenly here were the media in his face asking, in effect, how it felt to be dead wrong. Well, thanks a lot, but Mark Schoeberl wasn't exactly ready to be declared wrong. Yet NOZE-1 had spoken, he was told, and chemistry exclusively was at work.

"You send chemists to Antarctica, of course they're going to trash dynamics," snapped Schoeberl in a quoted reply.[5]

Dr. Linwood Callis of NASA's Langley Research Center, the man who had come up with a solar-driven ozone loss theory, was equally upset. "Their suggestion that the solar cycle is not playing a role in this is wrong. And even if it's not wrong, it's certainly premature."

"Aha!" proclaimed one broadcast reporter with thinly disguised glee. "We have a world-class controversy!"

Whether there was or wasn't a real war, the need to draw distinct battle lines and report opposite sides with the appropriate casualties and emotions and cliff-hanging mysteries of who would win and who would lose began to harden and sharpen the anger and upset of various "camps," raising the level of debate out of the scientific to the level of personal acrimony and bitter resentment. All this was unnecessary, but when Susan Solomon and her team stepped back onto U.S. soil after their successful mission to McMurdo, the unexpected cross fire that greeted them begged the question of how the controversy had started. Instead of accepting praise for gathering their important data, they found themselves dodging bullets for reporting it.

CHAPTER
5
Showdown at the CFC Corral

Pend Oreille Lake, Idaho, August 1986

So much for the vacation.

Dr. Jim Anderson replaced the telephone handset and stood a second on the front porch of his old family homestead, staring down at the green Coleman table—oblivious to the cool mountain breeze quietly brushing past Pend Oreille Lake and gently ruffling his prematurely white hair. His mind was elsewhere for the moment, racing from the conversation with NASA's Bob Watson all the way to circuit boards and lasers, through impossible deadlines and incredibly tight schedules. Was there really any way they could make it? Or had he just accepted an impossible task.

Fishing had been his only agenda item this morning, but that would have to wait. Perhaps for as much as a year or two.

The call wasn't a complete surprise. In some ways it was inevitable. He and Bob Watson were essentially thinking in harmony about the Antarctic ozone hole and what was going to be required to prove beyond reasonable doubt that chlorine—and CFC's—were the root cause. If that *was* true, there would be ClO free radicals to find in the Stratosphere. And Jim Anderson's superlative research group—the sharp, skilled corps of people he had assembled at Harvard—were unequaled in their knowledge of how to detect individual free radicals in measurements so minute the unit of measure would be parts-per-trillion.

Jim looked up at last, drinking in the peaceful vista of the lake, the majesty of ponderosa pine lining the shores and the backdrop hills and mountains of northern Idaho. His grandfather had purchased this place in 1912, before Teddy Roosevelt shoved the National Forest Act through Congress. They had been too long coming back here. He had promised his family this relaxing vacation—promised to make a month-long hole in his busy schedule at Harvard, and had finally succeeded. And now ...

The technology of the instrument package they would have to develop in less than a year was not a problem. His group had designed and built far more sophisticated instruments for their balloon experiments: packages of lasers and electronics controlled by computers and launched in thirty-five-thousand-pound payload packages almost as large as the Lunar Lander—packages carried to a hundred and thirty thousand feet.

But this one would have to fly on an aircraft—interface with the electronics and the mounting requirements of an aircraft flying above sixty thousand feet and able to penetrate the Antarctic vortex, searching out the evidence they might use to shortcut and defuse the growing scientific squabbling over the cause of the hole. There was only one aircraft that could go that high and fly that far: NASA's ER-2, a modified version of the Lockheed U-2 reconnaissance airplane.

His wife was looking his way now, questioning, knowing instinctively that the peace and quiet and relaxation they had sought in Idaho had probably just been shattered. The wheels were turning in Jim's head. She recognized the look.

Jim turned to her at last, explaining that Watson wanted him to build a totally reliable instrument to fly on the ER-2. He had approximately nine months and four hundred thousand dollars—while continuing his full schedule as a chemistry professor and leader of his research group and their other ongoing projects.

"How fast can you do it?" Watson had asked.

"It'll be a long shot to get it done by next summer," Anderson told him. And that was probably an understatement. There would be great difficulties in getting everything to fit, and even more of a challenge making it work right the first time—and every time. There could be no partial data from this. It had to be exactly right. The fact that a major thrust of the 1987 airborne expedition might well be riding on his shoulders was painfully apparent.

"What did you tell him?" she asked.

There were immediate phone calls he would need to make, wheels to put into motion, schedules to figure out, and travel plans to arrange—as well as a family to coordinate.

Jim Anderson looked up and smiled broadly, the incongruous nature of the tranquil setting and the frenetic assignment too amusing to ignore.

"I told him we'd do it."

As he canceled the remainder of his vacation and returned to Harvard, the frustration of the previous years began to melt away in the recognition of what Anderson and his Harvard team of graduate assistants, postdocs, and students might be able to accomplish. Bureaucratic bungling and mismanagement at the research center in Palestine, Texas—plus an interruption in the supply of the thin plastics needed for his high-altitude balloon probes—had retarded their balloon launch program since 1983. He had taken pride in his 1977 success at being the first to find hard evidence of ClO in the atmosphere (which proved to be a milestone in the battle to control CFC's, as well as important validation of the Rowland-Molina theory), but his research team had been increasingly on the sidelines since then. The enormous challenge of his new assignment from Watson would put his research team back in the race and give them a major opportunity to contribute.

The search was on for a smoking gun to explain the origin of the ozone hole. Even as Susan Solomon and the NOZE-1 team were arriving in New Zealand for their trip south to McMurdo Sound, Bob Watson was working overtime to piece together the major international airborne expedition for the following year—an expedition that had overall a single goal: verify the cause of the hole. Watson knew he could get instruments to measure almost every molecular suspect in the Antarctic atmosphere except two: ClO and bromine. Even before the NOZE-1 team had been fully assembled, it was obvious to most everyone that if the cause of the hole turned out to be chlorine, ClO in large quantities in different locations would have to be detected without question to prove that CFC–borne chlorine was responsible for the hole. If ClO *did* turn out to be the smoking gun, Bob Watson's people had to be equipped to detect it from the air—from within the chemically perturbed stratosphere itself—during the summer of 1987. Otherwise, the entire expedition could fail because the data could be considered incomplete. He couldn't take that risk, of course, which meant that Jim Anderson—the only one with such world-class expertise—had to be cajoled aboard.

Peter Wilkniss of the National Science Foundation had helped take the daily logistics and coordinating tasks for NOZE-1 off Bob Watson's shoulders by late April. Wilkniss had moved mountains to get Susan Solomon's group launched toward Antarctica in August, but for Watson, NOZE-1 was just the first challenge.[1] During the very same period, the daunting task of arranging a major expedition for the following summer of 1987 was looming large in Watson's mind—and on his calendar. He had to assume that Solomon and NOZE-1 would return empty-handed from McMurdo—it was the safest assumption to make, since any number of things could prevent them from getting enough data from which to draw even preliminary conclusions.

The summer of 1987, though, had to be different. He would have to assemble and orchestrate an army—the foot soldiers and air force of science—enough brainpower and instrumentation to go south and gather all the facts needed to answer the major questions: Is the hole natural or man-made; if man-made, are CFC's responsible; if so, could the ozone hole spread to lower latitudes, directly threatening population centers? Someone had remarked to Bob Watson that his 1987 expedition plans were already resembling a miniature Manhattan Project—a major commitment of the brightest minds and best equipment for a bar-none effort to solve a particular scientific problem that potentially threatened mankind. The characterization was right. Setting up the airborne mission for August 1987 would mean a breathtaking race filled with logistics nightmares and funding headaches that in many respects resembled the race to build an atom bomb in 1944–45.

In any event, they really had no choice. If the hole was man-made, it was a threat to its makers, especially if the governmental institutions of man failed to understand the problem and take immediate action.

Of course, if the hole was natural, the scientific community might look a little silly for launching such a frenetic campaign. But whatever answer came from the expedition, it would be the result of unified international effort, and international cooperation between science and government is a noble goal in itself.

Bob Watson knew the atmospheric scientific community could make no mistakes with the ozone hole emergency. He was already having nightmarish thoughts about the reaction of government and industry if they made the mistake of saying that the Antarctic ozone hole was caused by CFC's, only to be proven wrong. If the hole was later found to be a natural phenomenon, the CFC producers would shake their collective heads in disgust and say: "You guys were wrong about the ozone hole, and now you're saying the ozone layer is in trouble? Bull. You're not credible." The inevitable backlash from such an overstatement would throw a blanket of warm, fuzzy apathy over the policymakers, industry, and the public from which the scientific community—and the Earth's ozone layer—might never emerge. And the chances of getting true international cooperation in limiting CFC's after such a debate would be nil.

The stakes, in other words, were about as high as they could get. The credibility of everything that had been done on ozone depletion from Rowland-Molina to NOZE-1 could be blown away by one premature pronouncement, and Watson was determined to see that didn't happen.

Bob Watson carried the burden of such realities quietly and well. For years he had honed a growing ability to serve as a wise catalyst between science and policy, trying to find the best ways of relaying the true dangers of global ozone loss to the hearts and minds of those who had the power and position to do something substantive about it—without becoming too much of a target or a radical himself. He had learned to present the latest scientific facts and options in ways that even the skeptical chemical industry (and the irritated European countries) could respect, and one of the crowning achievements of his consensus-building efforts had been the NASA–UNEP report on atmospheric ozone released several months before in January 1986—a process that had begun in 1980 when Watson first came to Washington.[2]

The Beltway—Washington, D.C.—was a far cry from the Queen Mary College of London University where Watson had earned his Ph.D. in chemistry in 1973. Watson's interests in gas phase chemistry had led him across the Atlantic to become one of Harold Johnston's postdocs at Berkeley in 1973—right in the middle of the SST wars that were pitting Johnston against the world. Before that point, Bob Watson had regarded atmospheric chemistry as just another one of many scientific subdisciplines, and even though he had done his Ph.D. thesis on bromine and chlorine chemistry, not a word of it related directly to the atmosphere.

But the Berkeley experience threw him into the air, so to speak, and after moving to the University of Maryland in 1974, he began focusing his research projects on chlorine and bromine gas phase reactions which would later prove to be atmospherically significant. In 1975 he joined Caltech's Jet Propulsion Lab in Pasadena, California, where he set up his own research group, meeting the ozone layer nose-to-nose for the first time. Once he had discovered the subject, Watson—with the encouragement of JPL's leaders—began to take his communicative skills and enthusiasm into the public arena with outside talks and review articles that slowly but surely laid a foundation of effectiveness in dealing with thorny issues. In 1980, Bob Watson was loaned to NASA in Washington—while still on the JPL payroll—where he set up shop as head of the NASA Stratospheric Research Program.[3] NASA had finally recognized the need to broaden its research base in atmospheric problems, and Watson's Ph.D. had been the right one in the right place at the right time.

After the FDA and the EPA banned the use of CFC's in aerosols in 1976, Congress directed NASA to report back every other year on the health of the ozone layer, and Watson inherited that responsibility as he came aboard. But he quickly realized that NASA could fulfill that requirement less divisively than it had in the past. Issuing a careful, scholarly update on the ozone issue under NASA's name gave the effort an air of Yankee superiority, and in Europe there were those who considered that akin to American propaganda—the "American Position" on ozone. Since the United States (after unilaterally banning CFC's for aerosol use in 1978) had kept trying to force Europe against its will to do the same thing, the European Economic Community had long regarded North America's pressure with resentment and hostility—and their data with suspicion.

There was, however, a way to cure that international rivalry.

In Watson's view, too many reports were being issued each year by too many different panels and research bodies around the world, each of them saying different things and predicting ozone loss at differing rates. One report would peg future ozone loss at 40 percent, another at 3 percent, and yet another at 5 to 7 percent, and industry learned quickly how to pick their favorite ozone report and wave it fervently in the air to support their position. If you regarded an eventual 5 percent loss as acceptable, there was sure to be a report that would support that prediction. The built-in variations and contradictions of such a system gave those who didn't want to find an ozone problem—those who wanted the CFC problem to go away quietly—all the ammunition they needed to simply cite the paper that had the lowest figures of ozone loss. In 1979 alone, for instance, there were no fewer than six international reports.[4]

The National Academy of Sciences had added to the problem in late 1979 by once again altering its predictions of eventual global ozone loss—this time to 16.5 percent—and although the Environmental Protection Agency in the dying days of the Carter administration proposed new rules to limit CFC's (so-called Phase Two of EPA regulation, Phase One having covered aerosols in 1978), Watson could see that the nonstop scientific bickering was going to hamper the process of getting anything done to protect the ozone layer indefinitely.

Watson decided to widen the scope of the NASA report by inviting the World Meteorological Organization into the process along with the other "alphabet agencies," the Federal

Aviation Administration (FAA) and the National Oceanic and Atmospheric Administration (NOAA). That would make it an international effort *coordinated* by NASA—not *authored* by NASA. Sure enough, as soon as the joint effort was announced, the upcoming report lost its parochial veneer as the "American Governmental position," and doors began to open to the NASA-led scientists—doors that would have remained closed before. As a result, the ecumenical group turned out a high-quality report that predicted a 5 percent worldwide total ozone loss (to the National Science Foundation's 3 percent in their next report). The figures did nothing to stem the idea that the CFC-ozone problem was no longer a world problem, but it did set a precedent for international cooperation, and that in itself would prove invaluable just a few years down the road.

When it came time to do the next report in 1984, Watson was determined to make it even better by widening the scope again. This time he would include the United Nations Environmental Programme itself, along with the European Commission, the German Ministry of Science, and NASA, NOAA, and the FAA. The goal was the production of a truly international report that scientists and governments all over the globe would trust—and they achieved exactly that. The resulting document became *the* definitive report for 1985: A large international effort with 150 scientists from more than a dozen countries. Even the National Academy of Sciences/National Research Council recognized the quality of the Watson-orchestrated study and promptly abandoned their scheduled 1984 report. The NASA, Watson-led effort was a brilliant method of welding a consensus among potentially quarrelsome factions across international lines, and one that tried very hard to eliminate the politics and the policy questions and focus exclusively on the science.

And, it provided the policymakers with the science stated in ways they could use—a practice Watson would soon elevate to an art form.

Watson presided over the release of the massive effort in January 1986, just before jumping headlong into the NOZE-1 and NOZE-2 challenges. In two thousand pages the report—which was on the state of the atmosphere in general and not just on ozone—confirmed that CFC-borne chlorine was in fact destroying ozone, that damage was already occurring, and that what had once been only theory was now virtual fact and a major threat to the planet, the full scope and depth of which the scientific community could not adequately predict. By the year 2050, it said, the Earth's ozone layer would be depleted by anywhere from 4.9 to 9.4 percent, and *that* range of loss would only hold true if the rate of CFC emissions didn't exceed the 1980 rates.[5]

"They don't, and they won't," replied the chemical industry.

"They already have," was the response from Sherry Rowland and others.

The Reagan administration had compiled a tawdry record on the environment with such environmental pariahs as Anne Burford (EPA administrator) and James Watt (Secretary of the Interior), but with the departure of Burford's EPA replacement—William Ruckelshaus—in January 1985, a new EPA administrator named Lee Thomas came aboard with more environmental awareness and concern than anyone in authority at EPA since the days of Jimmy Carter. Thomas had the attitude that the EPA could do something substantive on the global ozone depletion problem. But he also understood that to be effective, whatever was done would need full international cooperation. A ban or cutback by the United States alone was next to useless from the point of view of the ozone layer. The other nations of the world that produced and consumed CFC's *must* agree to at least some controls.

And for that drama, the stage was already set.

In 1972 the UN Environmental Programme had been created as a loose framework, but the hope was that within that structure specific formal arguments could be concluded, actually limiting whatever needed to be limited. The United States had unilaterally banned CFC propellants in spray cans back in 1978, but few countries had followed suit. It wasn't until 1980 in Oslo, Sweden, that ozone issues finally gained a true international forum. There, Canada, Sweden, West Germany, the Netherlands, and Denmark joined the Norwegians in agreeing to wait a while longer for still more scientific proof before placing formal limits on CFC emissions worldwide.

In the end, the Oslo meeting ended up largely symbolic, despite the fact that it added a framework for future talks. The delegates were brave enough, however, to caution all the nations of the world that they had better begin reducing the volume of chlorofluorocarbons from any source. But predictably, European hostility killed hopes for any such formal limits.

In 1983—as all but a few began to regard the CFC problem as a thing of the past—Sweden, Finland, and Norway (who would have a lot to lose from ozone losses in northern latitudes) submitted a "draft protocol" for a worldwide ban on aerosols and controls on all uses of CFC's. It became known as the Nordic Annex and gathered the immediate opposition of the European Economic Community. Yet, over the next few years, thanks to significant changes in the negotiating positions of the United States (caused by the departure of Anne Burford and pressure from responsible environmental groups such as the Natural Resources Defense Council and the Environmental Defense Fund), the international community slowly lurched toward the eventual signing of the Vienna Convention in March 1985—a framework treaty that affirmed the signers' intentions to protect the atmosphere but left the details of specific limitations and methods to future negotiations and protocols.

Basically, the Vienna Accord was an agreement to continue negotiations, and in the view of many scientists it was a full decade behind schedule. Nevertheless, it provided at long last a mechanism to bring about global cooperation—a "Living Agreement" in the words of U.S. negotiator Richard Benedick—that could enable rapid international action when the evidence was finally clear.[6]

It was into the middle of this minefield that the new EPA

administrator Lee Thomas ventured in 1985, as Joe Farman was waiting for publication of his seminal paper about the ozone losses over Halley Bay.

The ozone issue was dead—wasn't it? The CFC industry and chemical manufacturers in general had genuinely convinced themselves that other than rearguard actions (by rapid environmentalists trying to stir up trouble) and with no substantive evidence, the problem had all but gone away. But now in the spring of 1985, here was an EPA administrator under *Reagan,* for crying out loud, who wanted to pick a fight and start the controversy all over again. The atmosphere became fractious immediately.

Thomas was convinced that ozone depletion was a valid issue that had a solution. More important, he became convinced that policy decisions to limit CFC's were appropriate even in the absence of scientific certainty—and the startling revelation of the Antarctic ozone hole propelled those feelings into the January 1986 release of the Stratospheric Ozone Protection Plan (the long-awaited EPA Phase Two of ozone controls). The way Thomas saw it, if the Vienna Convention wasn't going to produce a quick protocol to limit chlorofluorocarbons, the EPA would do it for them—at least in the United States—and would get the ball rolling by accelerating research efforts and calling new international workshops to seek international protocols under the Vienna framework. Since the newly discovered ozone hole over the South Pole was really there and totally unexplained—and since it had to be on the minds of everyone associated with the issue by the spring of 1986—it had to have influenced the process. But the EPA, worried about the same possibility of natural causes that bothered Bob Watson, made no formal mention of the hole. At that stage, international action against CFC's would have to be based on what they already knew, not on a galvanizing scientific mystery that might turn out to be ozone death by natural causes.

The industry reaction to the accelerating level of activity became one of "run with the crowd but keep firing over your shoulder." At the same time that funds from such industry groups as the Chemical Manufacturers Association were helping with the NOZE-1 expedition, other industry spokesmen were complaining that CFC's were the victims of unfair accusations. And when Sherry Rowland once again disregarded personal risk and announced quietly but firmly that he felt heterogeneous chemical reactions involving chlorine and nitrogen compounds were enabling CFC's to attack ozone in Antarctica, the response was a reprise of 1974: "That's completely fanciful! Where's your proof?"

The phrase "scientific uncertainty" gained slogan status once again, and industry scientists began circulating papers and reports that tried to show that CFC's were so important to mankind that a small loss of ozone protection was a reasonable price to pay to have them. Even the most pessimistic reports, they said, called for less than 10 percent ozone loss in the next fifty years, and that was "acceptable" for the incredible convenience of chlorofluorocarbons—especially since they couldn't be replaced.

Or could they?

More than a decade had passed since the major CFC manufacturers had started looking for substitutes for CFC-11 and CFC-12 (the most destructive types), but when the issue looked dead—and no one seemed to be making any moves to ban CFC's—the industry quit talking about the search for alternatives. Now, suddenly, the CFC manufacturers were faced with potential EPA and international bans. But with no substitutes ready for market, industry-generated scare stories were circulated that CFC replacements would drastically increase the expense of all CFC-related items, and would probably require abandonment of efficient car air conditioners as we know them. "There *are* no available, affordable substitutes!" one industry representative howled.

But the truth—which would embarrass Du Pont, for one—was simply that the CFC producers had stopped looking six years before.[7]

By June 10, 1986, the climate was ripe for confrontation—and Senator John Chafee of Rhode Island was going to provide the stage. As the NOZE-1 team raced to get ready for an August departure, Bob Watson, EPA officials, and many others concerned with global atmospheric issues trooped to the Hill to talk about ozone depletion—and the new worry about the greenhouse effect, which Chafee threw onto the table as he opened a session of the Environmental Pollution Subcommittee (of the Senate Committee on Environment and Public Works):

> Today we are beginning important hearings to discuss two related problems that stem from pollution of the Earth's fragile atmosphere: . . . ozone depletion; and . . . the greenhouse effect and climate change.

Congressional hearings sometimes have little to do with convincing lawmakers or their staffs one way or the other on important issues, but they provide a formal forum in which the most prominent players in any ongoing controversy can lay their case into "the record"—especially the lawmakers themselves.

> Ozone depletion and the greenhouse effect can no longer be treated solely as important scientific questions. They must be seen as critical problems facing the nations of the world.

And, if the press is there, hearings give people the opportunity to present their views to the general public in a setting that often exudes a sense of theater.

> This is not a matter of Chicken Little telling us the sky is falling. The scientific evidence, some of which we will hear today, is telling us we have a

problem: a serious problem. There is much that we know. There is a great deal that we can predict with a fair amount of certainty... But we will always be faced with a level of uncertainty.

As Senator Chafee began, there was a general feeling that with an expedition about to head for Antarctica in response to the appearance of the "Ozone Hole," the ozone issue could be approaching a climax, and media turnout was respectable—which was all the more important in raising the collective level of awareness of the public on the other atmospheric issue that had yet to really rise to a level of prominence in the public mind: the greenhouse effect and global warming.

There is no controversy about the existence of a greenhouse effect. The fact that there are certain gases that absorb the infrared energy radiating back up from the surface of the sunlit Earth, thus trapping some of the heat that would normally dissipate into space, is well known. A runaway greenhouse effect is what keeps the temperatures on the surface of Venus many hundreds of degrees hot by trapping the majority of the heat transferred to the planet by the sun. Carbon dioxide, CO_2, is the primary natural greenhouse gas on Earth, but there are others as well, such as N_2O (nitrous oxide), CH_4 (methane), and other so-called trace gases in the atmosphere—*including* chlorofluorocarbons. In fact, in addition to being an ozone assassin, each stable CFC molecule also happens to be more than fifteen thousand times more efficient than CO_2 in absorbing and retaining Earth's reflected infrared energy.

If the greenhouse effect is accelerated, Earth's average temperature will rise. This too is relatively undisputed. Increase the amount of greenhouse gases, trapping more of the heat radiated from Earth's surface in the form of infrared light, and eventually the average temperatures will have to go up. The concept of the greenhouse effect causing global warming then, is not in and of itself controversial.

And, the fact that **Earth's atmospheric gaseous balance has been significantly altered by human activity in the last hundred years,** is also beyond serious debate. The amount of carbon dioxide, for instance, has increased 25 percent over that period of time—a figure atmospheric scientists have derived by studying the amount of CO_2 trapped in air bubbles frozen in ice deposited prior to the Industrial Revolution, and comparing it to present-day samples.[8] We also know there were no chlorofluorocarbons before the twentieth century, and therefore: **The millions of tons of CFC's now in the lower atmosphere add to the greenhouse effect and have in and of themselves changed the gaseous equation.**

Well then, if no one's arguing the existence of the greenhouse effect as a valid theory, or denying its ability to cause global warming when the greenhouse gases are increased in percentage, or disputing the fact that humans have increased greenhouse gas percentages significantly, where's the controversy?

The actual *effects* of global warming. The debate is over what will—or will not—happen to Earth's biological and physical environment at the surface as a result of global warming, *and*, over whether global warming has already become detectable in climatic changes.

We may have more heat retained in the atmosphere if global warming is a reality, for instance, but will more clouds appear as a direct result and counteract the heat by filtering more sunlight out and thus limiting the inbound solar heat? Or could there be other natural checks and balances to prevent our raising the average temperatures of Earth?

Even if Earth's average temperature *does* go up significantly, will we really change the climates of many locations, raising average sea levels, and causing a host of other horrors? ***That's* the controversy: whether there will be an effect, how bad it will be, and if the process has already begun.**

In the hearing of June 10, 1986, there was no question in the minds of the knowledgeable participants whether from government, industry, or science, that the greatest problem was deciding exactly *when* science had provided enough evidence—proof—to justify real action in cutting back either on CFC's or changing energy production methods in order to reduce CO_2 emissions. To the eyes of the environmentalists and the cynical, the chemical industry and rabid antienvironmentalists simply turned their heads and ignored the issue, buying additional time in which to make more profit by tossing off the justifications that "the science is too uncertain," "we don't know enough yet," and "we need more research before we imperil entire industries on the basis of emotional reaction." And at the bottom of all such resistance was the concept that chemicals released into the atmosphere should not be considered harmful and dangerous until *proven* harmful and dangerous.

On the other side of the argument was the legitimate concern of industry and government that passing laws with major economic impact without adequate evidence could damage the leadership position of the United States in the field of industrial chemicals, if it were discovered later that the move was unnecessary. How many times, industry scientists had asked repeatedly in congressional hearings, has science reversed itself on "findings"? Suppose we ban CFC's, destroy the industry and many thousands of jobs, then discover that CFC's weren't harmful after all? Suppose we change our energy production system, stop burning coal, build more nuclear plants to keep from increasing carbon dioxide emissions, and ban nitrogen-based agricultural fertilizers, only to find the greenhouse effect causes global warming that is adequately absorbed by natural planetary weather systems and the oceans?

Assumptions, in other words, were the key. Should we assume that what we don't know *can't* hurt us? Or should we take action on the assumption that it *can*? Said Chafee:

To my mind, the risks are so great that we must avoid continuing on a path that will irreversibly alter our environment unless we know that it is safe to proceed down that path. Scientists have characterized our treatment of the greenhouse effect as a global experiment. It strikes me as a form of planetary Russian roulette. We should not be experimenting with the Earth's life support systems until we know that when the experiment is concluded, the results will be benign. As Russell Peterson, former chairman of the President's Council on Environmental Quality ... has said, "We cannot afford to give chemicals the same constitutional rights that we enjoy under the law. Chemicals are not innocent until proven guilty."

The protocol in congressional hearings is usually a sense of cordial decorum and gentlemanly conduct mixed with nearly obsequious compliments between subcommittee or committee members all compressed in time under the intense pressure of House and Senate calendars. Senators and congressmen arrive and depart in the middle of the proceedings. Their aides busily whisper updates on what's happening, the questions they're scheduled to ask the witness when their turn comes, and reminders when it's time to rush from the room for another committee hearing or floor vote. In the midst of all the decorous confusion, at least one member of the committee or subcommittee keeps the questions and proceedings going. In this case, the job fell to the chairman of the subcommittee, John Chafee, who was concerned and knowledgeable about both the ozone and greenhouse problems, even though his Republican party had been tarred with a broad brush as being a sworn enemy of the environment.

Six of the thirteen members of the Environmental Pollution Subcommittee were present on the morning of June 10—which was unusual—and as the first panel (a group of witnesses who all sit together at the witness table, give their statements individually, and then answer questions collectively) took their seats, those journalists in the room who had been following the ozone or greenhouse effect/global-warming issues realized they were looking at an Olympian league of scientific players: Dr. Sherwood Rowland of the University of California at Irvine, Dr. Robert Watson of NASA, and Dr. James Hansen of NASA (who had pioneered the contemporary warnings about greenhouse/global warming).

Bob Watson spoke first:

> There should be no doubt today that there is compelling evidence that the composition of the atmosphere is changing at a rapid rate on a global scale. All of these changes are caused by human activities; that is, combustion and agricultural policies ... [and] ... CFC's are of anthropogenic origin ... We must stop thinking of these issues as isolated issues and consider them to be coupled both scientifically and politically.

Bob Watson was acutely aware that, as usual, he was walking a thin line. He could point out the political and policy pathways, but he couldn't lead anybody down those paths. He could say, "If you want to reduce ozone destruction, then you must take the following steps." He could say, "There are dangers to society in doing what we are doing." But he couldn't say, "NASA recommends you take the following steps immediately," or even, "Here's what I suggest you do in making policy judgment." The difference between his role and that of Sherry Rowland, as Watson had learned to perceive it, was in taking that final step—making the positive policy recommendation. That's where science stopped and activism began.[9] Thank God a man of Sherry Rowland's stature and conscience was around to carefully cross the boundary. But a professional peacemaker—a scientific catalyst, which was Watson's role—

had to constantly search for that dividing line, and make certain he didn't cross it, and thus limit his effectiveness. His statements to Chafee's subcommittee reflected that caution:

> I believe global warming is inevitable. It is only a question of the magnitude and the timing.
>
> It is not wise to experiment on the planet Earth by allowing concentrations of these trace gases to increase without full understanding of the consequences.
>
> A ... logical approach, *if you want to ban or regulate* the fluorocarbons, is to put an emissions cap on the total amount of gas that goes into the atmosphere [emphasis added]. It doesn't matter what the use is. It matters that it's getting into the atmosphere. I think [an international] protocol at this time that limits the amounts of fluorocarbons that get into the atmosphere is an extremely wise approach ... the area of disagreement, is exactly where should we put that limit?

Not that Sherry Rowland, by contrast to Watson, wasn't surgically careful about what he said. But having become very frustrated at the inability of Congress—whose members had listened during the early eighties with glazed eyes and abject disinterest to his warnings—Rowland had long ago realized that scientists' greatest difficulty in dealing with the policy-making and political process was simply in making themselves clear. He had become experienced at doing just that, but there were precious few colleagues who had the same skills—even if they possessed the same courage.

When it was Rowland's turn, he pulled the microphone closer, and looking over the top of his half-frame reading glasses, fixed John Chafee with a steady gaze.

> The first paper which Dr. Molina and I published in June, 1974, carried the outline of our

theory that the chlorofluorocarbon gases would eventually produce serious depletion in stratospheric ozone. ... With the benefit of 12 years of intensive study, [that outline can] now serve equally well as a brief summary of the facts of the chlorofluorocarbon-ozone problem.

The Antarctic ozone hole has arrived as a profound shock, first because the losses of ozone are massive; and second, because it was completely unprecedented... We are now in the position of having chosen to tolerate some unspecified amount of ozone depletion, and are now wondering how badly we have miscalculated. We now have a hole in the ozone layer which will last for a century or more, even if the world were to stop further emissions of chlorofluorocarbons today, which is, of course, impossible. Will the Antarctic hole deepen? Will it spread, and how soon, to other latitudes in both hemispheres? Can we afford to go for another 5 to 10 years of wait and see, of measuring and monitoring and study? ... the obvious answer is to discontinue this experiment without waiting for all of the answers.

Sitting quietly to Sherry Rowland's right, Jim Hansen was still uncomfortable with such appearances, though he understood them. It was the accelerating concern over global warming that had placed him in the witness chair—that, and his pivotal research work into the issue. It took time away from his work, but it was part of being a NASA research team leader.

Hansen was a quiet and studious scientist who had already looked into the abyss of the greenhouse effect, and come back somewhat shaken. His normal role as director of the NASA Goddard Institute for Space Studies was leading the small group of researchers in space physics at the Institute, a "think tank" somewhat incongruously placed above Tom's Restaurant in uptown Manhattan near Columbia University. Only once, about five years earlier, had he testified before Congress—that time to a House committee chaired by Congressman Albert Gore of Tennessee (who was perhaps the first member of Congress to grasp the importance of the Greenhouse Effect).[10]

Hansen's involvement in what was rapidly becoming a controversial issue had been increasing ever since. But as he sat at the witness table on June 10, 1986, he felt uneasy. He was convinced he had important new information relevant to the greenhouse story, specifically an analysis of observed global temperatures which showed the 1980s, so far, to have been the warmest decade in the instrumental record, and a computer simulation which showed the 1980s as years of measurable warming and the 1990s becoming even hotter.[11] Neither the global data nor the climate simulation had been published yet in the refereed scientific literature, and that meant that any statement Hansen made based on his unpublished data would have to be made very carefully, and with great caution. To complicate things further, the main recommendation he wanted to make had been censored out of his written testimony by the White House Office of Management and Budget in a prehearing review of Hansen's proposed testimony.[12]

With carefully laid strategy, Jim Hansen showed color maps of observed and simulated global temperatures as he pointed out that both showed a tendency toward higher values in the 1980s. He included all the appropriate caveats about the significance of the available results, which were admittedly preliminary.[13] He warned the panel that the single scenario considered assumed "business as usual" growth of greenhouse gases the world over, but that when the computations were complete, they showed a warming by the year 2050 that was sufficient to significantly change the average number of hot days in various locations. Washington, D.C., for instance, which in 1986 suffered an average of thirty-five days a year with temperatures above 90° F, by the year 2050 (according to the computer model) could have an average of eighty-five such broiling days per year. Hansen hoped the figures would raise a few senatorial eyebrows and help make clear that a global warming of several degrees, which might otherwise seem trivial, in fact implied dramatic regional climatic change. But the words were so cautious, the warnings so carefully couched in the proper caveats and conditions, that his words were unlikely to ignite any firestorm of public controversy.

In fact, the EPA had already taken the lead in some respects by issuing a little list of horrors that might be expected to occur if the greenhouse effect accelerated global warming in ways that couldn't be absorbed without significant climatic change worldwide: First on the list was what might happen to the Great American Prairie (as it was called in the nineteenth century), the region of the United States that became the "breadbasket of America" with modern farming techniques and irrigation water. With global average temperature rises of only a few degrees, it could become a near desert, with massive crop failures, complete collapse of the major multistate aquifers (the underground water-table systems fed by rivers and rainwater percolation from which wells supply much of the agricultural irrigation water). With increasingly milder winters and warmer summers, Alberta, Canada, could replace Kansas, Nebraska, and the Dakotas as the major North American wheat-producing real estate, leaving the Midwest a useless expanse of ruined, windblown, overheated land. A semi-arid region would become arid—and that's enough for agricultural collapse.

The temperatures humans would have to live with would not rise any more precipitously than the average global rise in most predictions, but the *frequency* of over-one-hundred-degrees days would climb dramatically in the United States, driving air-conditioning requirements and the need for electricity to unheard of levels.

Even more ominous would be the slow melting of glacial ice locked in the Earth's ice caps north and south, plus the heat-driven expansion of existing seawater, and massive heat-induced increases in evaporation rates in bodies of water such as the Gulf of Mexico (which could drive the frequency and strength of major hurricanes to unprecedented levels at the same time that global sea levels would be rising). Sea level increases of between two to twelve feet could occur by the year 2100, the EPA warned, and all it would take to flood low coastal cities such as New Orleans would be a three-foot average rise.

The effects of global warming, in other words, might be catastrophic.

Might be.

The truth was, no one knew for certain.

Jim Hansen was acutely aware that the climate models his research group were using were primitive compared to the dynamics of the climate itself. Even with the highest-speed computers available—such as the Cray supercomputer—such models gave them only a "dirty crystal ball," as one researcher put it. Even if the known parameters for last week's weather were programmed into such a model, it couldn't accurately predict today's.

Yet with all their limitations, the climate models were getting better, and they were giving some indications—telling some frightening tales—of the effects of global warming on North America. Hansen would tell Chafee and his subcommittee what the models indicated, but like Watson, he wanted to stop short of drawing conclusions he might not be able to defend—especially since the warming he thought had already occurred was still so small (on the order of one degree Fahrenheit) it couldn't be positively distinguished from the "background noise" of natural Earth temperature fluctuations. Hansen adjusted a slide on an overhead projector aimed at a screen to one side of the witness table.

> This is an estimate of when the greenhouse warming should begin to rise above the noise level; that is, the level of natural climate variability. . . . By

the 1990's the expected warming rises above the noise level. In fact, the model shows that in 20 years, the global warming should reach about 1°C, which would be the warmest that the Earth has been in the last 100,000 years.

The fact that CO_2 in Earth's atmosphere had increased by 25 percent since the Industrial Revolution was a well-publicized figure by 1986. Trace gases, too, were increasing, and both Hansen and Watson told the subcommittee that the trace gases alone contributed about one half of the global-warming effect (especially CFC's, which as greenhouse gases are many times more potent per molecule than CO_2).

The "King" question is an old standby in congressional hearings. "If you were king of the world, sir, what would *you* do?" It was hackneyed and somewhat comical—but it often elicited interesting answers. Senator John Chafee dusted the "King" question off once more when Jim Hansen had finished.

"Suppose each of you were king and you had what you might call unlimited authority. What would you do about this problem? I will start with King Watson," Chafee asked, a broad smile on his face.

"Thanks, once again," Bob Watson replied. "I think we have to look at all the gases, including methane, nitrous oxide, and CO_2. These are very hard policy options. I think if you isolate or focus on one gas, that is, fluorocarbons, which are relatively easy to regulate, that would be the wrong approach. We will be back here soon saying, 'We covered the fluorocarbons; now what do we do about the CO_2 and nitrous oxide?'"

"Dr. Rowland?"

"If I were king," began Sherry Rowland, "the first thing I would do is consult with the queen, who is sitting behind me." He let the laughter die down before continuing. "I would start on the chlorofluorocarbons . . . there are problems in controlling the other trace gases, [but] I think there is a possibility that we can control, in some respects, the emissions of carbon dioxide..."

"Dr. Hansen?"

"I am sorry if I sound like a befuddled scientist rather than a king," said Hansen, "but I would like to understand the problem better before I order any dramatic actions. It is a very complicated global system, and we are just beginning to be able to model it."

The problem of scientific certainty and scientific proof had been enshrined in yet another atmospheric issue, this one far more complex than ozone. Responsible scientists couldn't gloss over the uncertainties, yet policymakers, industry, and others were not going to be convinced to act until they saw the bullet coming. Of course, by then—as with the ozone layer—it could be too late to dodge.

In the ozone war, there was a single enemy: chlorofluorocarbons, man-made and replaceable.

In the greenhouse–global-warming arena, though, the enemy is everywhere: Man's production of energy—as well as his very breath—adds carbon dioxide to the atmosphere at ever-accelerating rates as the world's population accelerates in numbers. With the exception of CFC's, the gases involved are natural—only the quantities are man-made phenomena.

As Pogo once said, "We has met the enemy, and he is us."

For Bob Watson, June 10 and 11, 1986, were difficult days out of an impossible schedule. After his brief statement—and before Sherry Rowland took over—Watson galvanized the hearing room audience with a movie sequence of color shots of the Antarctic ozone hole developing and undulating through its life-death cycle in images taken by the TOMS satellite and pieced together in a time-lapse format.[14] But in his analysis of what was causing it, he stayed strictly in the middle of the road, carefully telling the panel—and the media watching him—that the cause of the hole was still a mystery.

But it was a mystery they were determined to solve. And

if not in the Austral winter and spring of 1986, then for certain in 1987.

Bob Watson had promised to throw everything into the airborne expedition of 1987. Balloons, satellite data relayed real-time to wherever they decided to headquarter, airplanes from NASA–Ames at Moffett Field, California, and more ground-based data from McMurdo. Since the theories were coming in with specific parameters that could be confirmed or blown out of the tub by actual observations, the challenge was to get the actual observations with sufficient quality that no one could seriously question the raw findings. That was a relatively easy goal to enunciate, but even for a peripatetic powerhouse it was almost too much to juggle at once.

"It can't be done that fast!" was a comment Watson was getting more and more used to hearing as he requested, cajoled, assigned, pulled, tugged, and shoved fellow scientists, bureaucrats, and politicians into line. With ten million in funding almost secured (including some funds from the Chemical Manufacturers Association) and growing public concern, NASA's support was no problem from the top down. Getting the aerial hardware, though, involved a specialized set of problems.

To start with, Watson wanted NASA's ER-2—a modified version of the famous Lockheed-built U-2 spy plane—to carry the experimental package Jim Anderson was going to build into the heart of the vortex and the ozone hole itself.[15] The demands on the ER-2's pilot were always high in physical terms, but this mission would present special challenges. Yet there was no other airplane that could do the job. Certainly the ER-2 pilots were used to wearing the uncomfortable pressure suits for many hours in the small single-seat cockpit (equipped with an ejection seat), and they had dealt with inherent worries about fuel lines freezing before, too. That was a normal operational concern for a craft that could fly so high in the stratosphere. The airspace above the Antarctic continent, however, is the coldest on Earth, especially during the

Austral night. If the fuel should freeze and cause a flameout, the pilot probably couldn't restart the turbojet above forty thousand feet, and the temperatures below would be below the freezing point of their fuel as well. That would mean a forced landing or bailout, and the prospects for survival over either the Antarctic or the oceans surrounding the continent were not good—especially since there was no room aboard to carry a life raft or other survival gear. A pilot caught in that position would have to depend on a search-and-rescue mission that could take many hours to arrive.

Even if all those worries weren't enough, NASA's flight operations people explained to Watson, the built-in crosswind limitations created by the impressive wingspan and centerbody landing gear arrangement of the ER-2 meant that the plane had to carry enough fuel to divert to airports as much as three hundred miles distant in case the return field had out-of-limits winds.

"You're a bundle of cheer!" Watson told them.

"Well, I can't say it can't be done, but it's difficult. We'd have a lot of questions. We'd have to know it's really necessary," was the reply from Moffett Field in California, the home base of the NASA atmospheric research fleet.

In a word, the NASA flight folks were underwhelmed. Nevertheless, the NASA fleet was there to study the atmosphere, and the thorough professionals who ran the flight program were not about to turn down a valid request for such an important mission without doing a full risk-analysis. As soon as Watson's request had been relayed by NASA–Ames scientist Estelle Condon, they dove into the task.

NASA's ER-2 was an invaluable tool. The airplane had been designed and built by the famed Kelly Johnson team at Lockheed's so-called Skunk Works aircraft design unit in California. It was the same ahead-of-its-time group that had given birth to the Lockheed Constellation, the F-104, and the recently retired and still-unbeaten speed demon, the SR-71 Blackbird.

By late July Condon had arranged a meeting between the NASA pilots and Bob Watson, and their risk analysis complete, they reported back: We'll do it, but the flight path has to remain within gliding distance of a land mass—something that could be accomplished without compromising Watson's mission. They were on! He had his basic platform, on which Jim Anderson's key ClO–detecting experiment package (along with many others) would be mounted.

The NASA DC-8 presented an easier proposition. Formerly used for over fifteen years as a plush intercontinental jetliner in the colors of the original Braniff International on its South American route system, the venerable old DC-8-62 was scheduled to be retrofitted with new engines and alter the interior for use as an airborne scientific lab capable of flying over eleven hours and nearly six thousand miles nonstop.[16] Watson planned to use the DC-8 as a manned observation platform outfitted with an additional array of sampling instruments that could fly through the vortex sampling the lower levels of the stratosphere.

It was the multiple data sets that would help establish the truth in one expedition. Even before Susan Solomon had unpacked and cut a hole in the roof at McMurdo, Watson knew that whatever she and the NOZE-1 team might find would be roundly criticized as limited data because it came from the ground only at one location. Even if Hofmann's balloons were successful, the single location was a major limitation. But in 1987, with the ER-2 at high level, the DC-8 at a lower level, the satellite coverage they would arrange, plus coordinated ground data from McMurdo, Halley Bay, and perhaps other locations around the Antarctic continent, they could preclude the argument that perhaps a limited sample had missed something important.

Jim Anderson had returned to Harvard from his Idaho cabin within days of Watson's call. He immediately enlisted the help of NOAA's Art Schmeltekopf (whose spectrometer was at work with Susan Solomon at McMurdo Sound) in designing his instrument for a high-speed airstream. It had to be able to detect chlorine monoxide concentrations as rare as one in ten trillion parts in an airstream flashing through the sampling probe at nearly 80 percent of the speed of sound. The aerodynamics of the problem alone were going to be a major challenge, and by the end of fall over a thousand blueprints of the probe would be drawn in conjunction with Lockheed, whose airplane had to carry it with complete compatibility.

Meanwhile, Estelle Condon of NASA–Ames took a team of her people south to Chile and Argentina to find an airport from which to launch the expedition. After much searching they chose Punta Arenas, Chile, a military-civilian field with a town nearby overlooking the Straits of Magellan at the south end of South America—an airport with one critical taxiway so dilapidated NASA had to send money in the next few months to repave it—a job that would eventually be finished mere weeks before the start of the expedition.

With most of the major tasks assigned, Bob Watson—who nevertheless continued to buzz through a daily schedule of responsibilities and contacts that would have brought a hyperactive teenager to his knees—began to turn some of his attention to a vital meeting coming up in Geneva in December. While he had helped to hold the lid firmly on any official conclusions about the incestuous relationships of ozone and chlorine and CFC's in the Antarctic night, the ozone hole was no secret to government or industry, and attitudes had been bending, changing, and generally undergoing an amazing metamorphosis during the summer and early fall. There was no question that the atmosphere at the international negotiations to begin on December 1 in Geneva was going to be very different from before. Just about everyone's position had changed, and those changes had begun several months before.

Back on the second day of the Senate Environmental Pollution Subcommittee hearings on June 11, 1986, EPA administrator Lee Thomas shocked the Reagan White House and put himself in instant political jeopardy when he departed from the administration's party line. The scientific uncertainty about ozone depletion, he was supposed to say, was too great to justify asking the CFC industry to bear the financial burdens of a cutback. While certain members of the administration had softened their attitude from the Anne Burford days of open hostility toward the environment and cynicism toward EPA's role in the equation, the specter of American corporations being forced to accept new government regulations from a president who was determined to get government off the backs of business had helped squelch anything but the ubiquitous call for "more research."

But Lee Thomas had gone home on the tenth, read his staff briefing papers and his prepared, White House-approved statement for delivery to Chafee's subcommittee, and had a heretical attack of individual thinking.

Thomas's prepared speech, as with those for most government agency chiefs, had been written for him, sanitized, and carefully checked with the administration overseers at the Office of Management and Budget (who ostensibly wanted to check such testimony to make certain recommendations are not out of step with administration budget requests). The Reagan policy of "do nothing to impact the business community without irrefutable proof" was clearly visible in some of the verbiage Thomas was expected to repeat before Senator Chafee's subcommittee:

> *On chlorofluorocarbons and ozone depletion:* Given the scientific uncertainties, we recognize that any action taken now has a cost associated with it which, as we learn more, may prove unwarranted. Thus, any analysis of whether actions to slow emissions [of CFC's] are necessary must compare the costs and risks of acting now or acting later.
>
> *On the international negotiations for CFC lim-*

itations: To the extent that action is needed, it is essential that the international community move forward to deal with these issues together. Further, we realize that our analysis must include all trace gases that may modify the ozone layer and not just CFC's. In the case of CFC's, we recognize that they are extremely important chemicals used across a broad spectrum of industrial and consumer goods. For some uses, we recognize that no effective alternative chemical currently exists.

Finally, we recognize that the potential risks we face are generally long-term and that, if action proves warranted, any regulatory approach selected should be structured in a way which minimizes costs and disruption to producers and users.

And on climate change due to global warming: Yet, without [this] knowledge of the possible implications of continuing to add greenhouse gases to our atmosphere, any decision on future policy action would be premature.

Thomas appeared the next morning before the Subcommittee and dutifully filed the prepared statement for the record, but largely ignored it in his brief verbal remarks. Skipping over the hard-core Reaganite rhetoric, he ended by tossing a personal incendiary remark in the record that would earn him a backlash from other sectors of the administration:

Clearly, decisions can't wait on certainty in the science; the global implications are too great, and I believe some intervention will have to be our course of action.

Predictably, the rest of the president's men—the Deputy Secretary of Commerce, the newly nominated White House science adviser and NASA chief William Graham, and witnesses from the Departments of Energy and State all followed the party line. Enacting regulations too soon, they all warned, was likely to create significant economic disruption and foreign relations problems.

At the 1986 CFC release rate of more than 700,000 tons a year, the time it took the Reagan soldiers to deliver their warnings against premature action spanned the additional release of another 7,777,800 additional pounds of CFC's, every pound of which will eventually reach the upper Stratosphere sometime within the next fifty years. It was ironic that after having taken the apparent lead in 1978 on protection of the ozone layer (by banning CFC spray cans), the United States was lagging behind in 1986, since the presence of the ozone hole—and the recognition of how long the CFC's in the atmosphere would literally be hanging around to threaten ozone—had caused a break in the logjam on international action.

Perhaps it was the looming presence of the Antarctic ozone loss hovering on the edge of each delegate's mind that propelled it, but attitudes began to shift. At an international workshop held by the United Nations Environmental Programme in Leesburg, Virginia, in July, CFC manufacturers from around the world finally admitted that production rates were in fact rising. It had been no secret that the 1980 production rates had been grossly exceeded for many years, but the Japanese and Europeans had maintained the fiction that there was no growth, arguing that if no more than 10 percent of the ozone layer was sacrificed with 1980 production rates, so be it—it was a small price to pay for human convenience and large profits. Now, however, in 1986, the companies that produced them, and the negotiators for their various governments, dropped the fiction. At the same meeting, the giant Du Pont company admitted for the first time that finding substitutes for CFC's wasn't really a problem. There would be a higher *price*, of course, but there would be substitutes available. With that, another fiction bit the dust.

The shocks were just beginning, however. The Alliance for Responsible CFC Policy, a virulent industry organization that for fourteen years had blindly fought Sherry Rowland and anyone else who would seek to connect CFC's to ozone loss, suddenly announced on September 16 that it would support governmental or international treaty limits on the *growth* of CFC's.[17] It wasn't a capitulation, exactly, but it was akin to the tobacco industry voluntarily calling for warning labels on cigarette packages. In dropping his bomb, the president of the industry lobby group, Richard Barnett, made sure to warn everyone that the change of position did *not* mean the industry really believed there was any real evidence that CFC's could harm ozone or cause a risk to humans, but they were taking the action simply because "the science is not sufficiently developed to tell us that there is no risk in the future." His citation of recent NASA depletion estimates for the future and the Antarctic "mystery" gave some heady clues to the true motivation. Apparently Barnett and his group had seen the train coming toward them, and decided to buy a ticket rather than block the tracks.

Within a week, Du Pont had called for a worldwide CFC production limit. Whatever Susan Solomon and her colleagues might find as they worked in the Antarctic cold, the very existence of the ozone hole had stunned the chemical industry into action.

The key, however, was getting a worldwide agreement to cut back on CFC production. In their defense, the Reagan administration hard-liners knew that unilateral American action would partially remove any incentive for Europe, Japan, the Third World, or any other CFC producer such as the Soviets to follow suit. If the United States *did* jump first, domestically imposing new limits, yet the other nations did not, the gesture would be worse than useless—it would actually *impede* the process of getting CFC's out of Earth's atmosphere altogether.

The next scheduled round of international talks under the

Vienna Accord were scheduled for December 1 in Geneva, Switzerland—a meeting that would once again try to move everyone closer to some sort of specific agreement to limit chlorofluorocarbons. But before the American negotiator Richard Benedick and his team of advisers (including Bob Watson of NASA and Dr. Dan Albritton of NOAA) could get to Geneva, the EPA under Lee Thomas's leadership joined (or pulled) the State Department into calling for a near-term freeze of CFC's (and the related halogens) at current levels, plus an eventual phaseout. Whether or not it had been approved by or had come from 1600 Pennsylvania Avenue, the American position had been announced.

In Geneva, the light and reason of the scientific message met the international brick wall of self-interest. The United States asked for the freeze and a 95 percent phasedown of CFC's by the year 2000, and the Japanese and European Economic Community turned the idea down flat. A freeze to current levels, maybe, they said. But a 95 percent phaseout? Forget it!

By then, however, NOZE-1 had returned to a controversy that did nothing if not highlight the possibility of a chlorine solution to the ozone hole. The effect of rapidly changing attitudes could be seen progressively as another meeting in Vienna in February resulted in some progress. In many minds, the final protocol signing in Montreal scheduled for September began to look like a realistic possibility.

By summer of 1987 the next-to-final form of that protocol had been written, environmentalists around the country and in the United Kingdom were winning media attention with anti–CFC campaigns that targeted everything from McDonald's use of CFC–blown food containers to solvent users, and Don Heath's satellite data suddenly came up showing something no one but Sherry Rowland quite believed could happen: a worldwide 4 percent drop in total ozone.

Bob Watson wasn't at all sure if he believed the figures. The satellite spectrographic analysis could be out of calibration and drifting, but the implications were too important to ignore, and with NOZE-2 and the airborne expedition about ready to depart, Watson formally convened an Ozone Trends Panel which he assembled from a group that had met earlier in December to consider Don Heath's satellite data. The Trends Panel thus included Sherry Rowland and many of the most respected atmospheric scientists on earth with the idea in mind that by doing an exhaustive examination of all available ground station data and comparing it to Heath's satellite data, the panel could make a reliable determination of what—if anything—they were seeing. If there really *was* a worldwide ozone loss—regardless of the final verdict of the Antarctic ozone hole—the need to get rid of all CFC emissions immediately would go up at least one order of magnitude.

A total of forty-three nations sent representatives to Montreal on September 14, 1987, and though a last-minute negotiating change by the United States almost torpedoed the protocol, by 10 P.M. the following evening all the attendees had signed the document—an unprecedented coming together of global proportions that agreed to freeze CFC and halon production and consumption at 1986 levels by 1990, a 20 percent cutback by 1994, and a 30 percent additional cutback by 1999.

More important, however, was the breakthrough clause that all the signatory nations would meet again eventually to eliminate chemicals that damaged or destroyed the earth's ozone layer. It set up a formal framework as well for calling everyone back for more stringent action if the scientific community finally appeared at the world's doorstep with the long-awaited smoking gun.

Bob Watson knew as he left Montreal that the failure of the Toronto Group to get a CFC use-limiting protocol several years before in Vienna had been a blessing in disguise. They had inadvertently avoided a significant mistake, because controlling particular uses was really a useless method of protecting the ozone layer. "Ozone," Watson was fond of saying, "doesn't care *where* the chlorine atoms come from, it will be diminished by them whatever their origin, whether from spray cans or air conditioners." To negotiate protocols on each and every specific usage would have taken decades as industry fought and scrapped over every decision, and other governments moved to protect their self-interests. But placing limits on production and consumption covered all bases, and that was exactly the effect of the Montreal Agreement. In this instance, waiting had paid off.

In Irvine, California, however, Sherry Rowland was not impressed. For fourteen years he had tried to bring the planet to its senses, and received everything from personal abuse to constant industry pleas for more time to "be certain the scientific conclusions are correct." In the meantime, over eight million additional tons of CFC's had been produced, and over five million tons had been released into the atmosphere. Montreal was a good step, *if* it were ratified (nations representing a minimum of two thirds of the world's CFC consumption had to ratify the treaty before it could become effective, and the process would take a few years at best). But it might yet be too little too late to save the ozone layer from major, sustained damage that could last fifty to seventy-five years.

"What you've been listening to," one respected scientist explained, "has been the noisy process of democracy in action determining policy."

Perhaps so. But as NOZE-2 and the army of 150 scientists entered their second month of arduous, tense work in Punta Arenas, Chile, on the day the Montreal Agreement was signed, the realization was in the minds of many who had watched the process ebb and flow that the "noisy process of democracy" might have already taken too long—and may ultimately have cost too much.

CHAPTER
6
Burden of Proof

Punta Arenas Airport, Chile, August 16, 1987

Jim Anderson hesitated a moment outside of Hangar One and looked to the south—into the unseen maw of the Antarctic ozone hole they had come to probe. The wind of a South American winter whipped his hair as it blew in from the cold seas of the Pacific and across the small, twisted peninsula of land on which the city called Punta Arenas—Cape of Sandy Gravel—was built.

The NASA ER-2 was out of sight now within the confines of the hangar, most of its experimental packages, including Jim Anderson's, already locked in place within its body or mounted on the wings. The single-engine Lockheed craft fairly bristled with scientific sampling devices, and every one of them was considered vital. If the current theories about the cause of the hole should turn out to be wrong, they'd be forced to go back to the raw data for clues. In the end, it could be some obscure scrap of information from something as simple as a temperature probe that led them to the real answer, which was why Bob Watson and Jim Margitan of NASA had made the decision early on that no data—and no scientific package—was to be considered superfluous (though there was a go/no-go list of which packages could be inoperative and still allow a single flight).

Jim Anderson looked to the south once again, in the direction of the town, which sat on the western side of the famed Straits of Magellan, looking across toward Tierra del Fuego. The airfield sat at approximately latitude 53 degrees S, and beyond the Punta Arenas, Tierra del Fuego area there was no more land for 650 miles across Drake Passage until Antarctica.[1]

It was hard to believe that wooden sailing ships had battled the elements to sail down the Straits in transit from the Atlantic to the Pacific and vice versa, but for centuries sailors had done just that, many of them coming to grief in the clutches of gale-force winds that swept this desolate part of the world and drove fragile ships onto the lethal rocks.

In terms of distance from the Equator, Scotland and the upper United Kingdom and most of Ireland are farther toward their respective pole than Punta Arenas is toward its pole. But the south end of the planet is unique. The winds blow harder near and around the Earth's southern latitudes than in the north because of the deeper cold of Antarctica and the southern ice cap, and the cold temperatures blowing in Jim Anderson's face on the evening of August 16 simply added to the threatening nature of the cold seas just to the south, around Cape Horn, the southernmost landfall of the continent.

The two expedition airplanes would soar easily over those cold and turbulent waters in the weeks ahead, or so they planned. The ER-2, which had arrived the day before, could be forced to stay on the ground by high winds, but the retrofitted NASA DC-8 should have no trouble sticking to the flight plan. The converted airliner was still parked on the ramp back at Moffett Field, California, fussed over by NASA maintenance technicians getting it ready for its flight south on the nineteenth, ready to carry experiments that would complement the extensive package of instruments aboard the ER-2, instruments that included Anderson's main ClO and bromine sampler. The ER-2 would actually fly *into* the hole taking *in situ* measurements at altitudes between fifty-four thousand and sixty-eight thousand feet, while the former Braniff jet was scheduled to cruise *below* the level of ozone destruction at twenty-nine to forty-one thousand feet, its numerous experimental, observational packages of optics and lasers looking up into the hole, with high-speed data streams flowing onto magnetic tapes and into the computers that would analyze them after each flight. The territory below the big jetliner would be just as forbidding and cold and hostile and remote as that traversed by the ER-2—especially when flying over the heart of Antarctica itself—but there were four engines to the Douglas beast and plenty of range from the more than one hundred thousand pounds of jet fuel carried in its wings. After all, in its earlier incarnation, Braniff ship 1809 had routinely departed Los Angeles, California, in the late evening and arrived in Santiago, Chile, the next morning after handling eleven hours in the air over the Pacific. It was a true intercontinental aircraft.

It was the ER-2 that would fly in harm's way, pushing its single-engine airframe to altitudes above sixty thousand feet over forbidding territory and seas that would offer no safe haven to a pilot with a major mechanical failure or a flameout. The similarities between the raw exposure of an ER-2 and the vulnerability of any eighteenth-century wooden sailing ship at first seemed too obvious to ignore, but the impressions were misleading. Neither was that vulnerable when operated by careful people. The ancient mariners were hardened, experienced men who knew what their ships could take and how to calculate the risks and lay to for months if necessary until conditions were right—and safe. So, too, would the Antarctic Airborne Ozone Expedition keep its ER-2 on the ground if the weather data, the mechanical condition of the aircraft, or any other factor wasn't within carefully calculated guidelines. Even when the ER-2 was airborne, there was no reckless, swashbuckling we-sail-into-danger aspect to the operation. The flight plans had been carefully drawn to keep the bird within gliding distance of land whenever possible, following routes that would take it parallel to Palmer Peninsula, for instance, on the way to the Antarctic continent itself. While chief pilot Ron Williams and the three other experienced ER-2 aviators weren't completely comfortable with the mission, they felt they had the risks within acceptable tolerances.

The ER-2 portion of the expedition was to be a series of ten flights, and they planned to start very conservatively, with the clear understanding that the whole thing might not work. Even after a year of frantic preparation and the expenditure of over ten million dollars of research money, the scientists had to accept the fact that if the ER-2 couldn't safely be pushed into the ozone hole over Antarctica—if the fuel lines were too close to freezing or the outside temperature dropped below $-130°$ F—that was that. The pilot would turn around, no questions asked.

Jim Anderson pushed that thought from his mind. Twelve months of Herculean effort would be totally wasted if his instrument didn't get a chance to taste the ozone hole in person.

Nearly a full year of nonstop tension and effort by every member of his handpicked research team had gone into the box of carefully designed but never assembled parts Jim Anderson had brought to Moffett Field in California in late May. They had drawn and redrawn a small library of blueprints during the previous months, always working in conjunction with Lockheed to make sure everything fit.

It was still a loose box of parts that Anderson began assembling at Moffett Field in late May as startled members of the NASA-Ames facility looked on. The expedition left in three months, they thought to themselves. How could Anderson possibly be ready? The damn thing wasn't even *together*!

But when screwed and bolted, riveted and latched, it fit the electrical system, the aerodynamics, and the physical mountings of the ER-2 as if it had been designed to factory specifications—which, in effect, it had.

But would it *work*?

True, as soon as it was assembled, the NASA pilots had flown it off the coast of California during the summer, soaring into the stratosphere and turning on the instrument. It not only worked, it actually detected some ClO, a startling finding in the mid-latitude atmosphere. And yes, they had checked and rechecked every connection and circuit to exhaustion before flying to Punta Arenas. But they had never been able to test-fly it in temperatures of $-120°$ F, and there was no assurance it would work in such conditions. As Anderson and almost every other scientist knows, what works in theory doesn't always work in practice. And there's always Murphy's law to dictate that what can go wrong, will.

Yet in Punta Arenas it had to work. As Jim knew so well, his challenge wasn't just to add to the data, his instrument was crucial to the expedition. If it couldn't detect ClO and BrO *inside* the hole at the right altitude and at the right times, the expedition wouldn't have a definitive answer, regardless of what other experiments worked.[2]

That, then, had become the prime directive: Come back with a definitive set of unquestionably good data from which an unassailable set of answers could be derived.

Several other members of the expedition were ready to go back to the Hotel Cabo de Horno, and Jim acknowledged them. There would be a 4 A.M. wake-up call, after which the first scheduled flight of the ER-2 would begin—*provided* NOAA's Adrian Tuck (project scientist for the expedition) and British meteorologist Peter Salter could look up from their postmidnight study of satellite-relayed weather data and give them the green light.

To the south, unseen and brooding, the Antarctic vortex waited, the stage for massive, sudden ozone destruction already set above sixty thousand feet, the end of the Austral winter and the first rays of sunlight penetrating only the perimeter.[3] But the chemical soup was primed, and (as would soon be discovered) the air was already denitrified—the chlorine-reservoir nitrogen compounds helplessly locked away in the ice crystals of the Polar Stratospheric Clouds, which lay out there somewhere in the darkness.[4] The mission to Punta Arenas was exciting and stimulating, but there was an occasional twinge of apprehension. Almost every member of the 150-person team had thought similar thoughts in the previous days of arrival and preparation. The ancient mariners facing the unknown dangers of the trackless sea to the south must have gazed into dark clouds and cold winds and had similar feelings of awe—tiny human beings challenging forces so much greater than themselves, and so much beyond human control.

But in this case, they were going to probe a dark atmospheric monster perhaps created by man himself.

Just before midnight on August 16, in the silence of space, the Nimbus-7 satellite passed some three hundred miles over Punta Arenas, at more than seventeen thousand miles per hour on polar orbit number 44,492, its Total Ozone Mapping Spectrometer (TOMS) dutifully making its routine observations of backscatter light from molecules of ozone in the atmosphere below. Minutes earlier, the TOMS unit had recorded the continued growth of a minihole in the ozone over the base of the Antarctic peninsula less than two thousand miles to the south. The hole was only a precursor to what was to come, but it was the first indication of unusual ozone destruction—a symptom that would grow by mid-September to Brobdingnagian proportions.

The Nimbus-7 data, in the form of binary digits, flashed down as radio waves to one of several dish antennas at Goldstone, California, then bounced back 22,500 miles to a NASA deep-space relay satellite and back down again to the Goddard Spaceflight Center in Greenbelt, Maryland, where the data stream flowed into a computer at Building Three. The latest transmission was stored on magnetic tape and manually taken to another building for a series of decoding and gridprocessing operations before the refined images were sent back skyward as more data bits to another communications satellite, and back down again to a dish receiver at Punta Arenas, where the data was assembled by computer into usable, near-real-time snapshots of the ozone picture over the south polar region.[5]

The lights had burned all night in the communications center in Hangar One at the Punta Arenas airport, the telephone lines and satellite relay circuits feeding Teletypes and facsimile machines, the ambience resembling a cross between a command post and an overcrowded electronics lab. Spread out on a table before chief project scientist Adrian Tuck of NOAA's Aeronomy Lab in Boulder was a cornucopia of data that had flowed down the same satellite link to Punta Arenas as the relayed TOMS images—weather information, charts, predictions, and readings transmitted by participating meteorologists a world away in England and Europe, all of it coalescing into the big picture that would lead to a singular go or no-go decision just before 4 A.M. Tuck, whose muscular physique, bushy eyebrows, and drooping mustache cut an image in direct contrast to his keen intellect and vital position as one of the key decision makers of the expedition, would make that call. There were strict criteria to meet before waking up the ER-2 pilot, let alone launching the valuable plane. The winds on the ground and aloft had to be right, the position of the minihole, the area of ozone loss, had to place it within reach, and the forecast had to be reasonable for the projected return time.

And by 4 A.M. on the seventeenth it had all come together: Go.

In the language of atmospheric science, the south polar atmosphere was not yet widely "perturbed" (largely changed from normal chemical balance)—loaded with chlorine and chlorine compounds, devoid of nitrogen compounds that might have tied up the chlorine, and basically ready to stage a repeat of the previous year's sudden ozone loss that had begun in early September, but not yet perturbed.[6] That would occur in full when the sunlight struck the chemically volatile mixture. But for now the stage was set, and that was the atmospheric brew the first ER-2 flights were supposed to sample as the growing daily levels of ultraviolet solar radiation marked the approaching end of the Austral night.

During the months of Antarctic winter darkness, the polar vortex winds circling the Antarctic continent effectively segregate the air within, over Antarctica, restraining it from freely mixing with the rest of the Earth's atmosphere outside, and forming a gigantic circular mass of very cold air measuring nearly four thousand statute miles in diameter![7] Some three hours after takeoff on each trip the ER-2 would fly through that invisible "wall" to sample what was inside: low nitrogen compounds, high ClO readings similar to those seen from McMurdo the year before, vast regions of polar stratospheric clouds, all accompanied by normal levels of ozone. If the chemical theory was correct—as Susan Solomon and her colleagues believed—once the sun had begun to shine, an army of chlorine molecules would be released to begin eating the ozone molecules, leaving greatly increased levels of ClO in their wake. Some cooling might occur with less ozone to absorb solar energy in the lower stratosphere, and that in turn might cause some dynamic movement of the air downward, but chemistry—and not dynamics—was thought to be the cause. And if Jim Anderson's instruments found ClO . . .

The minihole over the Palmer Peninsula was both a precursor and an anomaly, but the flight would try to navigate it.

Above Punta Arenas was a healthy ozone layer: approximately 300 DUs (Dobson Units). At the same time Palmer Station 720 nautical miles to the south had reported readings of 219 DU's by telephone link, while the Nimbus-7 was reporting levels over the rest of Antarctica in the mid-200s. Within a month there would soon be a 50 percent or greater drop in the amount of ozone within the vortex spread all across Antarctica—with ozone values less than 125 DU's—but for the moment the situation before polar sunrise seemed almost normal with the exception of Palmer Station.

As the team members began to place bare feet on cold floors at the Hotel Cabo de Horno, the meteorology group in Hangar One continued to watch the weather charts and ponder questions that only the upcoming flight could answer: Were the temperatures within the vortex too low for the ER-2 to safely handle? Would there really be polar stratospheric clouds in abundance (a key requirement if the chemical models were correct)?

And would the instrument packages work—especially Anderson's?

At 9:32 A.M. ER-2 chief pilot Ron Williams lifted his airplane off the Chilean runway and turned south, climbing steeply into a clear sky as Anderson and his team watched and worried and mentally rechecked everything. It would be some five hours and forty-one minutes before Williams returned—six long hours before they could plug in their computers and verify that the four-hundred-thousand-dollar package built into a bomb-shaped pod beneath the left wing had really worked.

It had worked flawlessly on the way south, but that didn't count. The ER-2 had sampled stratospheric air while flying from NASA Ames in California all the way to Panama, and then on to Puerto Montt, Chile, and finally to Punta Arenas. When it touched down, Jim Anderson had been pleased and relieved when the data transfer showed perfect operation.

But that was at moderate stratospheric temperature

ranges no worse than −85° F. Now Ron Williams was going to be poking the nose of the jet into temperatures lower than −128° F (or −89° C), and even the best of electronic instruments could develop problems in such frigid readings. There were scores of tiny electrical connections inside the instrument, all of them gold-covered, but even gold could develop tiny cracks that could break a circuit in extremely low temperatures, and such breaks were very difficult to find. Especially if they were intermittent.

More than an hour before takeoff time Ron Williams had begun breathing pure oxygen to get rid of the nitrogen in his body, a precaution against the bends. Getting into the pressure suit required helpers and time, and so did the preflight breathing. Without it, the lower pressure at high altitude even with a pressure suit could cause nitrogen bubbles to come out of solution in his bloodstream. The malady was the same that scuba divers had to guard against. The bends could take the form of anything from mild itching to complete nervous system collapse or respiratory failure.

The discomfort of a pressure suit becomes a moot point in a high-altitude, single-seat airplane. You can't live without it, can't get out of it, and can't do anything about it, so you simply put it out of your mind and live with it. Ron Williams had been flying U-2 and ER-2 missions for two decades, and for him there was a special rhythm and pacing to the cockpit duties during predeparture, takeoff, and climb that all give way to a more relaxed feeling at altitude.

Airline, transport, or even bomber pilots could take off the shoulder harness at cruise altitude, and perhaps even go back for a cup of coffee (after transferring control to a copilot, of course). But single-seat airplanes are their own world, and to an experienced single-seat airman, the rhythms and duties and routine at altitude become very natural (if not totally comfortable), the visual richness outside the canopy a familiar pleasure.

Except today.

WHAT GOES UP [157]

Ron Williams had expected clouds as he climbed to altitude, but not *at* altitude. Yet he was in continuous cloud layers, and they were the strangest types he'd ever experienced.

The altimeter had finally steadied at sixty thousand feet, far above the troposphere and well into the stratosphere—a region that normally contains no clouds, and no weather systems. True, in the mid and lower latitudes, major thunderstorms with incredibly powerful updrafts can blow clouds and precipitation through the tropopause and up to fifty or sixty thousand feet, but there were no thunderstorms here. Only continuous layers of what resembled cirrus clouds. But cirrus clouds only occurred at lower levels.

Ron craned his neck to look above him, trying to estimate the distance to the higher layers of clouds by judging their speed of passage. They seemed to go from sixty-five thousand feet all the way up to eighty thousand feet, as best he could estimate. That was unheard of!

Wispy clouds of ice crystals, gossamer and otherworldly, continued to pass his canopy in silence as he marveled at the fact that the models had been right. These were the polar stratospheric clouds (PSC's) but they were far more extensive than expected, and 450 nautical miles from the northern edge of the vortex. He had expected to find them, if at all, only inside.

Williams looked at the beauty of the passing veils of iced aerosols, aware that the theories predicted they were made of water compounds and nitric acid molecules locked within those tiny crystals. And their surfaces were suspected of being the butcher block of ozone destruction—the crucibles for the strange stratospheric chemistry that had taken everyone by surprise. In all his years in the stratosphere, Williams had never encountered such formations.[8]

The temperature was another shock as he neared the wall of the south polar vortex—the encircled body of atmosphere some scientists had given the controversial title of "containment vessel." The agreed-upon cutoff point for continued flight was −130° F. Below that, fuel icing could cause an unrecoverable flameout, with no warm air below to thaw the fuel lines and permit an in-flight engine restart. Williams had been watching the needle drop as he got close to the vortex. If it hit −130° F, he would have to turn around whether he had penetrated the vortex or not.

Working with the checklist, he had long since activated the experiments one by one before settling back, monitoring his bird, flying what was essentially a docile aircraft in a hostile environment of nonstop PSC's. Now he wondered what those instruments were recording.

Finally the ER-2 approached the wall of the vortex. Not a wall, really—more of a zone—a massive area of polar air restrained from free mixing with the air from the rest of the planet by the continuous clockwise movement of winds circling the Antarctic continent.

He couldn't see it, of course. It existed as a boundary of numbers and lines on the weather and navigation charts they had prepared for him. Williams checked his position again on the digital readout of the inertial navigation system and made a course correction. The temperature was still dropping, but he was getting close to the location of the lowest ozone readings—the minihole—that Nimbus 7's mindless robotic spectrograph had reported more than twelve hours before.

There was a small ozone meter in his cockpit, but no other indications of what the constellation of sensitive instruments he was shepherding were sensing. He had only the instrument panel lights to indicate that all electrical systems were on, and power was going to each appropriate location—such as Anderson's instrument pod slung under his left wing.

Was there a low ozone reading being recorded out there, and perhaps high ClO? Ron Williams found himself wondering.

As did Jim Anderson as the hours ticked by back at Punta Arenas.

The wait for the ER-2's return was an agony. There were

enough duties to keep everyone busy, but the hours weighed heavily nonetheless. When the word finally rippled through the hangar offices and the communications center that Williams was on landing approach, the news emptied the building.

Among their other worries had been the possibility that crosswinds could go out of limits and force Williams to take the ER-2 with all its experimental packages to a diversion field 720 miles to the north. If that happened, it would be three days before they could refuel and repatriate the bird— and three days before they knew what, if anything, had been recorded under the left wing.

Slowly, gracefully, the ER-2 approached the runway, touching down and slowing, the wing tips (supported on take-off by tiny outrigger wheels) kept level by pilot skill, the main bulk supported by the center-body landing gear. Williams taxied in and shut down, running through his checklists as the cockpit ladder was rolled out and a swarm of scientists and technicians converged around the aircraft.

Working with cold hands in the chilly temperatures, Jim Anderson and crew positioned a dolly beneath the left wing pod, winched the wooden cradle into position to hold the detachable pod, and disconnected the entire assembly, wheeling it back into the hangar where Anderson hurriedly opened one of the side hatches and connected the appropriate cords to his computer. When he was sure everything was ready, he triggered the on-board computer, asking it to transfer (dump) its recorded data.

Rows of numbers began to flash across the computer screen on the portable stand they had placed by the pod. If there were numbers, there was data, good or bad.

Suddenly the screen went blank. The data dump was still running, but there was nothing there.

Jim Anderson's stomach tightened as he began searching for the answer. The instrument had been working, it appeared, until the plane had reached the vortex itself. Then it stopped. One of his worst-case scenarios had come true: As the temperature dropped, *something* had opened up in one of the myriad circuits or connectors or wires.

As suddenly as it had quit, the data flashed on the screen again. When the temperature had risen outside the vortex, apparently, the connection had closed once more and the instrument had resumed sampling. There was good data outside the vortex, but it was all but useless unless it worked inside as well.

Jim Anderson was upset, but not crushed. He was determined to fix the problem but there was scant time. Adrian Tuck wanted the ER-2 to fly again on the eighteenth, which meant they had less than sixteen hours to solve the problem.

The other packages on board had worked the first time out (and had captured some of the most crucial data of the entire mission). But there was no question which one was the star of the show. And the star had stumbled.

It took an all-night session, but by the morning of August 18, with the ER-2 again cleared to fly, one of Anderson's team had written a computer diagnostic program that was to test the circuits one by one if the instrument failed again, recording exactly which of the gold-coated connectors had opened up.

Once again the ER-2 taxied out and took off, this time with NASA's Jim Barrilleaux at the controls. And once again when the airplane was back on the ramp hours later and the cables had been connected to Anderson's package, the data blinked out at the edge of the vortex.

But the diagnostic had worked. The culprit turned out to be the wiring interface between the on-board computer and the experimental package itself, and once identified, fixing it was simple. They had lost two flights, but hopefully the third—scheduled for the twenty-third—would gather a complete set of numbers, and the first full set of ClO measurements.

The DC-8 arrived on schedule on Saturday, August 22,

carrying additional team members, and on the morning of August 23—with the Nimbus 7 data showing Dobson values as low as 200 appearing along the edges of Antarctica—the ER-2 was cleared for its third flight. This time some genuinely low ozone values were in reach, just inside the vortex. Jim Anderson once again watched the launch with his team in a high state of anticipation. In many respects, this was the acid test at last.

It was nearly six o'clock and dark outside by the time the Anderson team once again huddled around a computer screen in Hangar One, checking the umbilical cable between the on-board computer in the airborne sampling package and the computer on a portable stand. Once again Jim Anderson punched in the appropriate command, ordering the on-board computer to transfer its memory contents, a process that began instantly, the familiar columns of figures dancing across the screen as they watched and waited. Within a minute the point would come—the segment that on the first two flights had blanked out as the faulty connection opened up, breaking the information circuit. Anderson was sure they had repaired it properly, but what if something else had opened up? There were only so many ER-2 flights possible, and only so much patience on the part of the NASA pilots. They couldn't be expected to push all the way to Antarctica day after day with a continuously faulty instrument.

The tension mounted as the seconds ticked by, Jim Anderson and his team realizing almost simultaneously that they were past the critical point in the data stream, yet the screen was still full of numbers and information.

The fix had worked.

Equally important, the instrument itself appeared to be providing good, steady data—and an emerging picture of a smoking gun.

The ClO concentrations had remained low and predictably rare from Punta Arenas southward until the ER-2 had penetrated the wall of the vortex along the Palmer Peninsula.[9]

Dobson readings had been dropping and rising from the low 200 range for the previous week, but when the instrument was flown inside the minihole, the ClO concentrations shot up—matching almost exactly the predicted correlation between rising ClO and dropping ozone concentrations. It was the first significant confirming data, and now with all systems ready and functioning and the DC-8 coming on line, they could make the final adjustments—get in position—to monitor the dramatic airborne aerosol drama that was about to unfold to the south.

CHAPTER
7
NOZE-II

McMurdo Sound, Antarctica, August 26, 1987

It was a homecoming of sorts. Even the hole in the ceiling was still there, as were many of the helpful and eager faces of the ITT Antarctica personnel Susan Solomon remembered so well from one year before. Of course the passage of twelve months hadn't added much to the ambience of her government-issue accommodations (and it wasn't exactly wonderful to hear on arrival that the pipes were frozen and water unavailable), but on this arrival there was no time lost wondering where things were, nor adjusting to the culture shock of McMurdo—the process that in 1986 had led to soul-searching questions, such as how being near the top of her profession had landed her at the bottom of the world.

This time they had hit the deck (or the ice) running, unloading and setting up in record time some twenty-five tons of instruments and supplies carried from Christchurch. There was an urgency to their efforts. The airborne expedition had already begun, a minihole had developed over the Palmer Peninsula, and the Austral dawn was getting close.

They were there to reaffirm and expand the 1986 NOZE-1 findings and to support the airborne expedition in Chile with even better ground-based observations than they'd managed before. In 1986 they had succeeded beyond their wildest hopes, and luck had been a factor in that success. In 1987, they were equipped to produce even better data.

And just as in Punta Arenas, there were no expendable experiments at McMurdo.

Susan and her colleagues from NOAA's Aeronomy Lab in Boulder had plotted their particular plan of attack with great care. In 1986 Art Schmeltekopf's spectrometer had measured three of what atmospheric science calls "spectral absorption features," but now they were equipped to look at several more and at different wavelengths. If the same data showed up—if there was still evidence of OClO in the atmosphere above McMurdo Sound—it would confirm their findings of the previous year. If not . . .

There was a certain tension in the air as they assembled the instruments and prepared to make the first observations, and as the team leader, Susan felt a special responsibility. After all, it had been their findings and preliminary conclusions after NOZE-1 that propelled the chemical theory of the ozone hole into greater prominence (despite the overblown brouhaha with Mark Schoeberl) and helped energize Bob Watson's efforts to get the airborne expedition off the ground figuratively and literally. With so much riding on accuracy, the data had to be right.

Susan's group had found OClO even in the daytime in 1986, and in roughly the predicted quantities. That—along with the 1986 readings taken in the moonlight which also showed OClO—had given them confidence that the OClO *was* truly there in significant quantities in the stratospheric polar vortex, and clearly an accessory-before-the-fact in ozone destruction on a continental scale.

Bob De Zafra and Phil Solomon also had worked painstakingly over their data after NOZE-1, laboring through several months of data evaluation that had finally yielded the prize: a measurement of greatly enhanced ClO, the kingpin molecule in the chlorine cycle. They too were back on the ice now, trying to verify their findings as well.

But what if Susan's findings on NOZE-1 had been wrong? What if the 1986 readings had been spurious, or misinterpreted? What if they returned and found no significant levels of ClO or OClO above McMurdo? It could mean the dynamicists were right after all, and the multimillion-dollar project already under way at Punta Arenas might be a totally embarrassing waste of time that Susan Solomon was partially responsible for creating.

Those scary thoughts had danced briefly across her mind as the airborne expedition got under way and her team prepared to depart for Christchurch and their second trip to the ice. The worry had abated with her hectic schedule, however—until now.

The Susan Solomon Rabbit's Foot Effect was working again, it seemed. The night sky of August 26 was crisp and clear overhead, and the bitter cold was a distant counterpoint to the excitement of having perfect conditions so soon after arrival. According to the weather charts, even the vortex had obligingly moved overhead almost on cue. Everything was set for the first night of data, but this time dancing with frostbite in the pale moonlight wasn't required: Having brought an automated tracker to the ice for their second visit, the team was able to sit safe and warm in the room below while moonbeams were automatically routed down the hole, and Schmeltekopf's spectrometer once again began digesting the electromagnetic radiation focused into its lens, breaking it to spectral lines and recording the results.

Now it was time to analyze the numbers, and Susan felt her throat tighten and her heart rate accelerate slightly as the final sequences were fed into the computer and she entered the keystrokes to start the analysis process.

It would take about three minutes for the computer program to compare the observed spectral data with the unique spectral signature of the molecules they were hoping to find—three minutes before the answer would pop out in the form of numerical values, giving her an initial idea of how much OClO was overhead—if there was any at all.

Susan tried to relax, but it was futile, and that was silly! Of *course* there was OClO up there this year. Of *course* the data from 1986—*her* data—would be confirmed. There was no way they could have spent two months finding consistent evidence of OClO a year before unless it was really there. And if it was there last year, it was there this year.

She looked at the second hand of her watch, the whirring and clicking noises from the computer sounding ominous as the machine accessed various tracks on the floppy disk drive and the internal hard disk.

Thirty seconds. That's all that had passed. Damn! Two and a half minutes to go. Thoughts of vindication and failure came in rapid succession, her apprehension growing.

It was always possible that the spectral lines had been misidentified somehow last year. Equipment failure. Human failure. The methods were sensitive and demanding, and you could screw it up.

Suppose that was the problem. Suppose they'd programmed Art's instrument or the computer incorrectly, and been zapped by the GIGO principle: garbage in, garbage out.

No. Not possible, she corrected herself. They had been too careful.

One minute thirty seconds. Halfway there.

Okay. The OClO is there. But suppose *this* method of analysis, these wavelengths, don't pick it up? We'll have to go back to last year's method and reverify. Or would a negative OClO result mean last year's method was wrong and misleading? Perhaps *this* one is incorrect. But Bob De Zafra found ClO last year, too. We couldn't both have been hoodwinked!

If it comes up negative, though, Susan thought, we're duty bound to report it, and that'll hit Punta Arenas like a hammer.

Thirty seconds left. Susan realized her apprehension was making the others nervous, too. Or perhaps they already were. She tried to adopt a nonchalant smile, but it wasn't working. They were all thinking the same thing, she realized: What if we're wrong? What if there's no OClO, or too little of it to be meaningful, let alone fit the models?

Thoughts of how she would inform Bob Watson in Punta Arenas that they were in deep trouble at McMurdo with inconsistent results almost tuned out the sound of the computer finishing its calculations.

The printer began operating suddenly, the tiny pins of the dot matrix mechanism frantically transferring numbers and letters to paper as Susan looked closely, peering at the emerging symbols, and understanding their meaning almost instantly.

"Yes!"

"What?"

"OClO! We got it!"

The numbers were almost the same as last year. They had used a different measuring parameter and come up with the same finding, which meant an extra element of control on their findings. They had been right in 1986, after all.

Deep down she had known that, of course. Intellectually. Logically. But not emotionally. The importance of the mission, the professional investment and risk, and the existence of competing ozone loss theories had caused her to doubt herself for a long and frightening moment. Those three minutes, which had seemed like an agony of days, were easily the three tensest minutes of her professional life.

But now they could get down to business.

McMurdo was the forgotten cipher in the 1987 ozone research equation as far as the media were concerned, but that meant nothing to the NOZE-2 team. This time they had even better instrumentation to read ClO wherever it might appear above McMurdo Station, and between Dave Hofmann's balloon launches and the impressive array of ground-based "remote sensing" instruments, the data sets began to grow almost as rapidly as the ozone hole itself.[1]

Joe Farman's Halley Bay had joined the reporting network on August 20, feeding daily Dobson Unit readings by satellite to Punta Arenas. McMurdo joined the net on the 27th, adding to ozone data flowing in from Palmer Station and by September 26, the American South Pole station (Amundsen-Scott Station), would also weigh in—a month behind the others, but an appreciated addition.

At Punta Arenas, the DC-8 took off on its first research flight on the twenty-eighth as the ER-2 made its fourth flight, the instrument packages aboard both aircraft working perfectly as they roughly paralleled each other's flight path to the base of the Palmer Peninsula, where the ER-2 turned for home while the DC-8 continued over the continent for five hundred miles before heading back. The minihole had undulated in and out of existence for the previous week near Palmer Station, but now the area of ozone loss was beginning to become coherent and quasi-stationary over the peninsula, with readings of Dobson Units in the low 200s.

Again on August 30 and September 2, both aircraft flew the same general routes, the DC-8 soaring all the way to the geographic South Pole before returning to Punta Arenas. The lowest Dobson readings had begun to develop on the opposite side of the continent, but the lowest values were still hovering above 200.

Until the morning of September 5.

As Adrian Tuck and his colleagues pored over the satellite charts, the TOMS figures, and ground-based reports on the morning of September 5, it seemed they were watching the curtain rise on the 1987 Antarctic ozone hole itself. Suddenly a small circle of TOMS readings around 175 had developed over the eastern edge of the Palmer Peninsula, and more than half of the continent was covered with values below 225. With the sun's rays striking farther and farther into the vortex with each passing day, the chemical fires had begun to burn, and as the chemistry and dynamics worked together, the levels of ozone were beginning to drop.[2]

They had a front-row seat, tracking the hole, sampling it from below and within, looking at it from above, sending balloons into it, and recording millions upon millions of bits of information that might take years to analyze fully, but that would eventually provide enough forensic evidence to convict.

On September 9 the lowest readings from within the drifting, widening ozone hole turned up values below 175 Dobson Units as the hole drifted south toward the pole. The edge of the vortex was now as far north as 59 degrees S, which meant that the ER-2's flight was the deepest penetration of the entire mission.

On September 11 the DC-8 flew to and below the area of the lowest ozone readings, after passing values higher than 425 Dobson Units outside the vortex (around latitude 50 degrees S). Once inside, however, the values plunged to less than 175 DU's for the majority of the flight. In fact, by the fourteenth, half the Antarctic airspace had dropped below 200 DU's, and by September 22—one day after the official end of the Austral night—the ER-2 on its twelfth and final flight pushed through the vortex wall and flew into a fully developed ozone hole deeper than that of 1986. Readings below 150 DU's now covered vast areas, and Jim Anderson's on-board instrument began recording ClO concentrations five hundred times higher than those found at comparable altitudes in the mid-latitude regions of the planet. As ClO went up, ozone went down—a strong negative correlation, and strong evidence.

By the middle of September the routine had become familiar—and wearing—but Jim Anderson was more relaxed than he had been for weeks. As the morning of September 22, 1987, dawned, his team was busy readying the instrument pod for the twelfth and last flight into the fully developed hole. The data for the previous eleven flights had been extraordinary, and nearly conclusive, although what they now knew had to remain a secret to the outside world for another week. Long before the expedition had started, the team members had agreed that there would be no daily announcements, leaks, phone tips, or any other possibility of dribbling partial information to the rest of the world that might cause embarrassing publicity and misinformation. By the end of the month, however, the cat would be out of the bag. Bob Watson would hold an end-of-expedition press conference, and as Jim Anderson knew, the findings would be startlingly conclusive on several major points. The expedition had already been a landmark success.

It was very gratifying to Anderson that, after the early problems, his instrument had performed so well, but some of that prideful edge was lost in the ominous meaning of the data. This was a chemical process, without question. Perhaps it was aided by dynamics—air movement—but what they were seeing was crystal clear: The force of sunlight in ultraviolet wavelengths broke up the molecular chlorine produced by the PSC's and delivered free chlorine into the Austral dawn to assassinate ozone, which could be read in high ClO and low O_3 values. The polar stratospheric clouds obviously set the stage, but the chlorine chemistry wrote the play.

Of course, the expedition would also be returning with a wealth of information on the polar stratospheric clouds, too, and that in itself was a scientific leap forward. Many of the other experimental packages had recorded extensive chemical profiles of the makeup of the PSC's, actually documenting the makeup of the frozen aerosols that set up the ozone-destroying reaction within those cloudy wisps of such ethereal beauty.

As the ER-2 climbed southbound for the last time, Anderson thought ahead to the joint statement of preliminary findings Bob Watson had arranged for the expedition to write and release to the world on September 30. Every day they had huddled behind closed doors to discuss the latest readings, carefully checking security badges and making certain no errant tape recorders or reporters were lurking to release partial data or misinterpreted opinions. Now, however, they had agreed to summarize the team's overall initial views, and the world reaction should be interesting. In almost everyone's opinion, the smoking gun had been found.

NOZE-2 Headquarters, McMurdo Sound, Antarctica

There were frozen gale-force winds howling outside in the Antarctic twilight, and the irritating static on the VHF radio made even the warm confines of their meeting room seem slightly strange—as if Susan Solomon and her group were watching themselves in an old black-and-white movie.

The DC-8 was overdue, at least as far as they were concerned. It was supposed to fly over McMurdo Station on its last research trip from Chile, this time bound all the way from Punta Arenas to Christchurch in one five-thousand-mile jump. They had looked forward to that—looked forward to talking with the pilots and perhaps some of their colleagues on board when the plane got close enough.

In a building a quarter mile distant, the aeronautical sector controllers of McMurdo were in contact with the NASA DC-8. One of the men in the room knew the NOZE-2 scientists were waiting for word of the plane as the voice of the pilots crackled through the radio speaker requesting a change in routing. There were unexpected winds on their nose in excess of one hundred knots, and McMurdo was now out of the question. Fuel reserves would stretch only so far.

The off-duty controller picked up the phone to relay the disappointing news.

Susan Solomon looked up at the ceiling and imagined the big jet out there in the lower reaches of the stratosphere, turning farther to the west and away from their position. It made her feel even more isolated, though they wouldn't have actually seen the jetliner anyway had it flown directly overhead.

It was a disappointment. The grand flyby, it seemed, had been canceled. Her NOZE team were partners in the Punta Arenas airborne expedition, but for the last month Chile had seemed so distant, the ER-2 and the DC-8 a world away. This final flight was going to close the gap a little.

They would be going home soon, too, a second successful expedition behind them, carrying huge volumes of data to be reduced and interpreted. The preliminary press conference and press release from Punta would come tomorrow, September 30, and they had been promised a copy by fax. Now, though, there was only the mental image of their friends a thousand miles distant at thirty-five thousand feet in a manmade bird, huge by human standards, but insignificant to the force of the wind and the rain that had kept them at bay.

On the morning of September 30, 1987, Bob Watson and Dan Albritton walked into the preplanned press conference at the Goddard Spaceflight Center at Greenbelt, Maryland, carrying copies of the so-called Fact Sheet bearing the logo patch of the Airborne Antarctic Ozone Experiment. When Watson had arranged for the end-of-mission conference, he had envisioned having the task of reporting divided results and unclear opinions, the intense debate of differing opinions ricocheting among the 150 scientists who would then be leaving Punta Arenas. They had all agreed that no one would talk outside of the common press release, which would include only those items that were clear and unambiguous, about which no one on the team would disagree.

Of course that could be a crushing limitation. There might be nothing of any significance that was "clear and unambiguous" when it was all over. Watson had wondered if he'd be facing a crowd of reporters and announcing that the only certain fact was that NASA aircraft had flown over Antarctica and returned with various readings too unclear to be reported yet. He might get booed out of the room, but that was a chance he would have to take. They must not repeat the type of disputed announcement and mixed signals that caused tumultuous infighting at the end of NOZE-1. That telephone press conference had been carried out honestly with the best of intentions, but there was no "clear and unambiguous" data to present. This time he planned to give a coordinated, carefully culled and manicured statement of undisputed findings.

Now the day had finally come, and a relieved Bob Watson had so much clear and unambiguous data to report, the press release ran to nine pages. This was going to be a pleasure!

The NASA end-of-expedition Fact Sheet contained almost everything the public might want to know. Watson was justifiably proud of what they had accomplished. Yet at the same time, what he had to report was darkly frightening for the planet: The ozone hole had reappeared and was even deeper than the year before; more than half the ozone had been destroyed over an area of roughly 12.5 million square miles of frozen land and sea, and chlorine was the cause:

> RESULTS AND THEIR RELATIONSHIP TO THEORIES
>
> The processes controlling the abundance and distribution of ozone in Antarctica are complex and intertwined. However, given the successful nature of this campaign, we are now in a position

to start to more fully appreciate the exquisite balance between the meteorological motions and the photochemistry. We will present our preliminary scientific findings as answers to a series of posed scientific questions that are relevant to public policy.

1. Did the springtime ozone hole occur over Antarctica in 1987?

Yes ... the abundance of ozone in August and September of 1987 was lower than any previous year at all latitudes south of 60 degrees. [Additional explanation]

2. Does the evidence indicate that both chemical and meteorological processes are responsible for the ozone hole?

The weight of the observational evidence suggests that both chemical and meteorological mechanisms perturbed the ozone. Additionally, it is clear that meteorology sets up the special conditions required for the perturbed chemistry.

3. Was the chemical composition of the Antarctic stratosphere observed to be perturbed?

Yes. It is evident that the chemical composition of the Antarctic stratosphere is highly perturbed ... [and] the present findings are consistent with the observations made last year from McMurdo. The distribution of chlorine species is significantly different from that observed at mid-latitudes, as is the abundance and distribution of nitrogen species. [Additional explanation]

Chlorine dioxide, $OClO$, which is most likely formed in a reaction sequence involving the ClO radical, was observed both day and night at highly elevated concentrations compared to those at mid-latitude ... [and are] ... consistent with measurements made from McMurdo last year. [Additional explanation]

4. How do the observed elevated ClO abundances support a chemical role in the formation of the ozone hole?

There is no longer debate as to whether ClO exists within the chemically perturbed region near 18 kilometers at abundances sufficient to destroy ozone *if* our current understanding of the chlorine-ozone catalytic cycle is correct. [Additional explanation and caveats]

[5 Through 8 omitted]

9. What are the global implications of the Antarctic ozone hole?

Until we understand the cause or causes of the spring-time Antarctic hole, we will not be able to address this key question in a responsible manner. Thus, at this time, it is premature for us to speculate on this important topic. However, as we continue to analyze the data that we have acquired and further test and expand the pictures that we have developed, we will be in a better position to address this important question.[3]

10. When will the data be in a form suitable for use in formulating national and international regulatory policies?

... The schedule for the assimilation and publication of the results is brisk. Peer-reviewed publications will appear in 1988. The results from the 1987 ground-based McMurdo campaign will likely appear on about the same schedule. Both sets of these completed conclusions would be the best basis for any possible policy reevaluations. The major international scientific review scheduled for 1989, which will serve as an input to the 1990 policy review of the Montreal Protocol, will have these conclusions available.

Susan Solomon finished the fax copy of the Goddard/Punta Arenas press conference and placed it on the table with a satisfied smile. The data were hanging together from all quarters, her team's efforts in 1986 had been reconfirmed, and there wasn't much more room for argument.

But there was an incredible amount of work ahead. They would be leaving McMurdo within two weeks, but the data reduction and ongoing meetings dovetailing with continued laboratory testing of the chemical theories would result in a virtual snowstorm of papers. There would probably be some sort of overall meeting in the spring of 1988, and she would be under great pressure to refine her data, conclusions, and theories in order to publish as quickly as possible. They felt confident that whatever the final truths about the chemical and dynamical processes going on over their heads, NOZE and Punta Arenas had the raw data to forge the final pieces of the puzzle.

In California, meanwhile, Mario Molina and his team were awaiting publication of the paper that would be one of the final key elements in understanding the strange chemistry of the Antarctic Stratosphere within the PSC's. With laboratory proof that reactions on the ice crystal surfaces of the PSC's could pull chlorine out of the safe harbor clutches of hydrochloric acid and leave it ready to convert to free chlorine in the first rays of sunlight—and with the breakthrough work on the existence and role of the ClO dimer (ClOOCl)—Molina and his team would end up an indispensable part of the expeditionary efforts (especially when their work was coupled with the ER-2 measurements of the effects of denitrification made by such team members as David Fahey and Ken Kelly—along with the work of Geoff Toon at McMurdo).

The scientific explanation might be close at hand, but no

one held any illusion that there were ways to repair the ozone hole anytime soon. As the TOMS satellite data had shown so clearly for previous years, by late November the drastically low readings of ozone would bottom out and begin to recover as the polar vortex began to break up and ozone-rich air from lower latitudes began to flow into the Antarctic air mass, mixing with the ozone-poor air and raising temperatures, the PSC's now gone, the air renitrogenated, and the high-speed ozone destruction halted. But the chlorine was still there, of course, with more coming as the millions of tons of CFC molecules in the troposphere continued to rise into the stratosphere until broken apart by UV radiation. There would be a rich supply of free chlorine for the next fifty years at least, and the ozone hole would probably grow larger and more pervasive each Austral spring when the sun's rays re-energized the ozone-destroying Antarctica. Understanding what caused it could accelerate the process of banning the CFC's which seemed to set it up, but decades, if not centuries, would be needed before the new annual phenomenon above the southern icescape would begin to reverse.

In the meantime, would the hole spread to more populated latitudes? It seemed unlikely, given the special frozen conditions of Antarctica and the PSC's, but the fact that ozone destruction had gone nonlinear and there had been virtually no theories or models to predict such a reaction still scared them all. What else might be out there in the mid-latitudes that could combine with a chlorine-loaded stratosphere and turn it into a raging bonfire of ozone destruction? Was there such a potential? Could another Krakatoa—a monstrous volcanic explosion throwing debris and sulfuric acid into the stratosphere—substitute for PSC's and start a nonlinear mid-latitude cycle of massive ozone destruction?

And what would be the response of the scientific community—not to mention the political world—if some poor atmospheric scientist made such a midnight discovery? What would be the response if another Mario Molina or Sherry Rowland concluded that there *was* something up there that could cause a mid-latitude ozone hole?

And then there was the Arctic. There were already bits and pieces of evidence from the TOMS satellite readings over the North Pole of an ozone deficiency there each spring. Somehow, Susan knew she would probably end up making a trek toward the north polar region as well. They couldn't very well ignore it. There were PSC's over the north polar region too in the winter, though not as many, and the Arctic vortex was weak and indeterminate compared to the southern version.

Yet, there were many millions of people in the northern latitudes. Norway, Finland, Sweden, northern Russia, Greenland, Iceland, Canada, and Alaska could be in grave danger if a northern ozone hole developed to the same extent as the Antarctic original. The thought of 50, 60, or 70 percent ozone loss and nearly unrestricted UV-B light shining down on human populations was frightening. UV-B caused cellular mutations and skin cancer. Even a small global ozone loss could greatly increase the number of skin cancer cases in the world, and the effects of massive Antarctic-style losses over the Arctic could be disastrous.

And meanwhile, CFC's continued in full production worldwide.

CHAPTER

8

The Cornfield Meet

Snowmass Village, Colorado, May 9, 1988

Sherry Rowland stood in the noonday sun slowly munching a bag of potato chips, his attention completely focused on Jim Anderson's words as they both leaned against the side of an ancient red pickup truck and talked about the momentous events of the past few months and the opening session of the Polar Ozone Workshop.

All around them other voices and forms and faces of the lunchtime crowd milled and congregated seemingly at random, forming small groups, then breaking away like collections of atoms broken from fragile molecules by the solar light and sent spinning toward new encounters. Some sat and ate their box lunches on the green grass of a cul-de-sac at the end of the driveway; others, mindful of the lobsteresque results of mixing fair skin and high-altitude sunlight for even thirty minutes, perched on brick walls and benches under the overhang of the Snowmass Convention Center, silently drinking in the beauty of the scene from the safety of the shade. There would be little time for beauty or recreation and reflection during the coming week, since most of their days would be spent in the nearby auditorium, discussing the fate of the latest addition to the Earth's endangered list: the ozone layer.

Jim Anderson shifted his position against the pickup as he finished a sentence, watching Sherry Rowland for a response and noticing the sunlight glinting with malevolent intensity off the reddening bald spot on Rowland's unprotected head. Both men had dressed casually, but now the khaki pants and blue sweater Anderson was wearing looked rumpled and a bit warm in the 60° temperatures, yet Rowland's tropical safari shirt seemed a bit out of place as well. After all, this was still Colorado. Only a month before, skiers had slalomed past this very spot, though now there wasn't a patch of ice or snow to be seen anywhere on the upper slopes. In fact most of the two hundred scientists on the attendance roster were in holiday attire, if not holiday spirits, and some were sporting newly acquired sunburns from unprotected rounds of golf at high altitude the day before. Snowmass had been picked as much for convenience and cost as the obvious ambience. May 9 through 13 was a dead week, the off-season lull between the winter ski season and the summer tourist crowd, and the perfect time for an isolated conference in a temporarily isolated place (which is scientific ambience in itself).

Isolated, that is, except for the army of workers preparing the ritzy resort for the next round of tourists. The constant whine of power saws and the hum of motorized vehicles accompanied the whisper of thin, high-altitude air blowing across the grassy ski slopes and through the collection of alpine-style condominiums, shops, and hotel—now totally commandeered by the scientists.

A few feet away from Rowland and Anderson, a reporter

waited patiently, watching carefully for an opportunity to approach the men. Only nine writers or journalists had shown up, along with two television reporters from CNN and one from the Canadian network CBC, but those who had made the trip fully realized the importance of the issues at hand, and they all understood the amazing fact that for the next five days, almost every senior scientist on Earth with any significant role in the ozone crisis was gathered in one place at the same time. It was an incredible chance to interview the movers and shakers of the issue.

The list of names was literally a who's who of the ozone wars: Sherry Rowland, Jim Anderson, Mario Molina, Joe Farman, Susan Solomon, Mark Schoeberl, Don Heath, Dan Albritton, Paul Crutzen, Richard Stolarski, Mike McElroy, Mack McFarland, Adrian Tuck, Steve Wofsy, and many more—all of them gathered under the wing of NASA and Bob Watson's direction for what had been stylized the Polar Ozone Workshop—a scientific gathering Adrian Tuck knew was going to turn into a "cornfield meet."

Tuck's efforts as a decision maker, meteorologist, and chief project scientist during the Punta Arenas expedition had been pivotal, and his role in the months afterward was even more so. His schedule through the fall and early spring was filled to overflowing with the struggle to bring the wide array of expedition scientists to the point of consensus and written papers concerning what they had found. Though Tuck's office was on the upper floor of NOAA's Aeronomy Lab in Boulder, he saw little of it December through February. The series of meetings held to define and refine the data and conclusions from Punta Arenas seemed nearly back-to-back, culminating in a pivotal three-day conference in picturesque Estes Park, Colorado, in February.

Tuck had hoped that Estes would be the final forum. There, according to his plan, he would be able to help nudge his colleagues into agreement, reconciling (at least for public consumption) everyone's differences and emerging with a unified view of the Punta Arenas findings as a singular set of papers, none of which would be diametrically opposed to any other.

Herding snakes would have been easier. The dynamicists and chemists simply could not get past their differing views of what the data really meant. Adrian Tuck needed to defuse any potential for having the upcoming major conference in Snowmass degenerate into real-time controversy over the exact method by which chlorine destroyed ozone over Antarctica each Austral spring. Yes, there was still legitimate doubt over the precise interactions of dynamic movement of the molecules in the Austral atmospheric soup and the role of such dynamics in the chemical destruction of ozone, but it was frustrating to scientists such as Jim Anderson, who had found the smoking gun of ClO. "We *know* what's occurring. We really don't need to know all the details of how it's occurring before acting. Those problems can be solved in time."

Tuck knew the scientific community would best serve the policymakers by providing a united front with their findings and conclusions—a state of theoretical unity—but such harmony wasn't going to happen. Even though the weather cooperated in Estes in cloistering the colliding scientists (freezing temperatures, wild winds, low clouds, and gloomy conditions kept everyone huddled inside the Inn of Estes Park along the frozen shores of Lake Estes), there was no way to thoroughly mollify the dynamicists. Yes, it *is* a chemical process, the dynamicists admitted, but it's set up by dynamical actions, and it's even broken up in late November and early December by dynamics, and there are some unresolved dynamic interactions involving heat flux of the Antarctic air mass and mixing of the polar air mass and the air outside the vortex that must be resolved.

"All we need right now are chemical explanations," said others, in effect.

By April, Adrian Tuck had effectively thrown up his hands. The barrel-chested, indefatigable world-class scientist from Britain who, with walrus mustache, incisive mind, and direct and friendly manner, had served as project scientist of the Punta Arenas expedition, knew that a peaceful consensus between the opposing camps was not in the cards.[1] "It's like the old days of railroading," he said, "when two trains, unable to communicate, would speed unknowingly toward each other on the same track through the middle of Kansas. Eventually, there was going to be a cornfield meet between the two—guaranteed to be messy."

And in this case, Tuck knew, in the high-altitude ski resort a few miles down the road from chic, eclectic Aspen, Colorado, there was, in fact, just such a cornfield setting: Snowmass.

Long before the inevitability of scientific conflict in Snowmass became apparent to Adrian Tuck, the apparent futility of providing adequate evidence to the chemical industry—and the policymakers who could force their hands—had become a frustrating, revolting reality to everyone involved with the expedition.

"What is it going to take to convince them?" The question had become almost a battle cry as soon as the team had returned from Punta Arenas and the reality had dawned that even their galvanizing preliminary findings weren't yet enough to retard CFC production worldwide. Nearly two hundred scientists in Chile and Antarctica, supported by hundreds of others on several continents, had spent ten million dollars and massive amounts of professional time in an unprecedented attempt to solve a matter of critical importance to the health of planet Earth, and they had succeeded. They had an answer, even if some of the precise details of the equation were still in debate. But by November 1987, there was no significant doubt among competent scientists that without man-made chlorine in the Earth's atmosphere, there would be no springtime hole in the ozone over Antarctica. With the Montreal Protocol, the Antarctic findings, and the growing realization that the same process of ozone destruction was probably occurring over the Arctic as well, it would seem logical that the CFC producers of the world—and *especially* the most responsible of all, giant Du Pont—would throw in the towel, dismantle their "tobacco institute"-style public relations body (known as Alliance for a Responsible CFC Policy), and admit that the products they were producing had now been tried and convicted of damaging our common atmosphere. Surely Du Pont would call for an immediate phaseout! After all, it was Du Pont that had hired a world-class, former NOAA scientist named Mack McFarland to keep them abreast of the latest scientific findings. McFarland, whose credentials and scientific honesty were beyond reproach, had even participated in the Punta Arenas expedition, and had already briefed Du Pont on the irrefutable evidence of CFC complicity that had been found in the stratosphere over Antarctica. Surely the giant chemical company couldn't justify waiting any longer, especially since CFC's were only a few percentage points of their overall business.

Yet when all the results were in and McFarland had briefed his employers that chlorine from CFC's was causing the hole in a strange process set up by meteorology, Du Pont's senior executives either never considered or flatly ignored the broader implications and pounced like a tiger on the unresolved issues between dynamics and chemistry, announcing that they would have to know *how much* of the ozone depletion was due to chlorine, and how much due to air mass movement, before a phaseout of CFC's would be "responsible" (a strange use of the word that sent several chemists to their dictionaries in puzzlement). It wasn't Mack McFarland's conclusion—it was the voice of profit-at-all-costs philosophy finally overriding good sense and responsibility. Du Pont had finally fallen from grace in the eyes of many scientists who had always felt more kindly toward their form of foot-dragging than toward the more radical defenders of CFC's who had wholly ignored the scientific realities from the start.

The events of fall and spring 1987–88 began to move with the speed of a musical comedy and the intricacies of a Kafka story. The Natural Resources Defense Council, one of the most respected environmentalist groups, had filed a lawsuit against the EPA in 1984 to try and force them to implement Phase Two regulations (Phase One being the spray can ban announced in 1978). Finally, in the fall of 1987—with the lawsuit closing in on them—the EPA acted, but its action was a bitter disappointment to many who also felt that the CFC restrictions of the Montreal Protocol itself were far too tepid. Instead of using the latest hard-won information from Punta Arenas, which was open and available to everyone in government by then, the EPA simply blinded itself to the new realities and adopted the Montreal Protocol limitations as if the Punta Arenas Expedition had never occurred. It was as if the United States of America—whose scientists clearly knew otherwise now—was stating officially that the Montreal Protocol's limitation on CFC production (only a 50 percent cutback by the year 2000) was realistic and adequate. The fact that the EPA laced its announcement on December 1, 1987, with gratuitous references to the *costs* (in terms of corporate profits and jobs) of switching away from CFC's was not lost on the environmentalists of the nation. The EPA, after all, was still an animal of the Reagan administration, despite the leadership of Lee Thomas as an administrator who seemed to be scientifically engaged with the problems at hand. And there was still the fervent belief in administration circles that unilateral action by the United States would do nothing to propel the rest of the world toward more stringent CFC regulation and phaseout.

The inherent, inadvertent fraud of Montreal was the idea that a 50 percent cutback (which was actually a 35 percent cutback when the additional production allowances to the Third World are considered) would somehow stabilize ozone destruction. For the EPA to continue to rely on such clearly fraudulent expectations was, in the uncharitable view, a cynical sham, and in the *most* charitable view, a significant mistake in international diplomatic strategy.

To the CFC industry, however, the EPA action was a relief.

In the periphery of the debate stood Bob Watson (the only key scientist of the ozone issue who would not be able to attend the Snowmass meeting he had orchestrated), Sherry Rowland, Mack McFarland, and numerous other top scientists who had come together at Watson's invitation to form the Ozone Trends Panel after Don Heath had made his startling announcement in 1986 that the TOMS satellite data from NIMBUS-7 were showing an overall worldwide drop in global ozone of 4 percent! If a worldwide 4 percent loss was a valid finding, it would be the first major evidence of global ozone layer destruction.

There were, however, serious doubts over the accuracy of that 4 percent figure.

NASA had been required by Congress to file its next two-year report on January 1, 1988, and Bob Watson (who had taken considerable pride in the fact that his agency had never been a day late with such reports) was faced with a dilemma: The Ozone Trends Panel results were to serve as the 1988 NASA report, and they weren't ready. They hadn't had enough time to thoroughly analyze the mountain of new data that had been turned up. "We could write something," Watson told the others, "but I don't think we'd be happy with it, or proud of it." With reluctance, he made the decision to slip the date three months instead, scheduling the report for a March release.

The Trends Panel had convened to analyze the TOMS data and determine its accuracy, because there was firm knowledge that the TOMS spectrographic instrument itself had degraded since launch, and was putting out erroneously low readings which were not being properly adjusted.[2] Since no one could climb into orbit and physically inspect the satellite (especially after the Challenger disaster of 1986), there was no way to find out directly. They would have to corroborate the data with readings from somewhere else, and there was really only one other control source: the worldwide constellation of Dobson observation posts known loosely as the Dobson network, which could give them a baseline to analyze (though it was too poorly coordinated to function as a true network).

But the findings from that network covering the last decade had already been carefully analyzed by professional statisticians, and there was no clear, downward trend in total worldwide ozone. Unless somehow they were misinterpreting the Dobson network, the satellite data seemed to be a direct and unresolvable contradiction.

The TOMS data couldn't be broken down any further, so the Trends Panel decided to take the Dobson network readings apart, using a method Sherry Rowland had already pioneered when he and graduate student Neil Harris had investigated the 1983 ozone readings decline at Arosa, Switzerland. Rowland and Harris had taken the Arosa Dobson station's data and compared *different months* as separate data blocks rather than homogenizing the readings year round. When they also looked at other northern latitude stations, the two chemists had been surprised to find previously unidentified wintertime ozone losses in the 1983 data (losses that Rowland began to suspect might be caused by the extra stratospheric debris inserted by the El Chichón volcano in Mexico acting as reactive surfaces for *heterogeneous* reactions).[3]

Maybe, thought the Panel, there are variations in the data that disappear when the months and the various latitude bands are all lumped together and averaged. That would turn out to be an understatement.

As the work of breaking the Dobson network data apart progressed, all the members—including Mack McFarland—were sworn to secrecy, and even Bob Watson refrained from telling his superiors and colleagues at NASA what they were finding—even when the findings began to get frightening. All of them would wait until the final executive summary was written and released. Watson couldn't inform NASA, McFarland was prohibited from discussing the trends with Du Pont, and Rowland couldn't publish. The limitations were necessary, but were a bit frustrating for some of the members, especially as December 1987 brought the reality that the chemical industry was going to keep dragging its heels over further CFC controls or phaseouts. With the industry thumbing its collective noses at the Antarctic results, the Trends Panel members by late January found themselves sitting uncomfortably on some very startling findings. There had been no consensus when Sherry Rowland presented his group's tentative conclusions to a meeting at the Trends Panel in Switzerland in December, but by late January and early February, the bombshell nature of the results was becoming an accepted reality. The Trends Panel in effect found themselves sitting on the very findings that would blow away the chemical industry's last line of defense.[4] It was tough to keep quiet until the formal release date, but they had no choice. Just as in the Airborne Expedition, the Trends Panel knew they had to have this one exactly right, thoroughly supported, and beyond scientific question by the time their findings hit the street—and the airwaves.

And in terms of global importance, their findings would be a quantum leap above those of Punta Arenas.

March 15 was rescheduled as the release date for the Trends Panel report, and Bob Watson found himself watching the clock, knowing instinctively that the debate that continued to rage through the winter season would be brought to a halt if only they could reveal what they already knew.

As Bob Watson wrestled with the decision to delay the Ozone Trends report in December 1987, an exhausted Susan Solomon was fielding a cheery call from John Meriwether, a physicist at the University of Michigan, whose field research had been based in far north Greenland, at the U.S.–owned Thule Air Force Base.

"Hey, wouldn't you like to come up here to balmy Greenland and see what's going on in the Arctic?"

Having planned a trip to Greenland as early as the previous June (before departing for McMurdo), Susan was instantly interested. Exhausted or not—instruments still unpacked and unchecked from McMurdo or not—Susan and her colleagues (George Mount, Ryan Sanders, and Roger Jakoubek) couldn't resist. By late January the atmospheric chemist who hated cold weather found herself once again wrapped up like a polar bear in an Arctic parka, this time near the top of the planet. The Arctic vortex—a weaker cousin of its Antarctic counterpart—had rolled over Thule the day they arrived, and by nightfall they were taking OClO measurements of the total column above the Greenland station.

There was one fifth as much OClO as over McMurdo, but that was still ten times more than normal. The temperatures at the appropriate altitude (30 millibars) overhead was in the $-80°$ to $-85°$ F range, which was just cold enough to emulate the Antarctic-style reactions in the polar stratospheric clouds, which were also overhead at Thule. Susan and her small team returned to Boulder in February, convinced that Bob Watson's intention to organize a Punta Arenas–style expedition to the Arctic in January 1989 was vital: "There is," she said, "evidence for anomalous chemistry taking place in the Arctic as well."[5]

But the embattled chemical manufacturers weren't in the mood to listen. As Susan's Aeronomy Lab team finally returned home to Boulder to stay and began unpacking their instruments at long last, a round of letters between Capitol Hill and Du Pont began that would demonstrate the depth of the major CFC producer's stubbornness—and leave a handful of senators virtually stunned.

Senator Robert Stafford of the Senate Committee on the Environment and Public Works had listened to testimony in an October 1987 hearing on ratification of the Montreal Protocol that deeply worried him. During the hearing, Sherry Rowland had once again trekked to the Hill to recommend immediate action to eliminate CFC's—action far beyond that of the Montreal Accord—while Mike McElroy of Harvard had called the exclusion of Antarctic Ozone Hole awareness from Montreal "a mistake," and the National Science Foundation's Peter Wilkniss had described the Antarctic ozone hole as so worrisome and dangerous that he was now afraid for the safety of scientists traveling there to study it! It was all too obvious to Stafford that the chemical industry had no defenses left and must throw in the towel. As with the many scientists who had reached the same conclusion, the senator just naturally assumed that a company such as Du Pont merely needed a slight nudge. The evidence was already overwhelming, yet Du Pont and the rest of the CFC industry had been stonewalling any change in their posture since late fall, despite what seemed to Stafford and others to be the inevitable end of CFC's. And for every day of inaction by Du Pont and the other CFC producers worldwide, thousands more tons of CFC's—laden with chlorine atoms bound irrevocably for the stratosphere and eventual ozone destruction—were being produced and shipped.

Robert Stafford rounded up fellow Senate subcommittee members Max Baucus and Dave Durenberger—all of whom had finally reached the breaking point—and suggested what form the "nudge" should take. As a result, on February 22, 1988, the three senators signed a letter to Richard E. Heckert, the chairman of the board of Du Pont, which reminded him of the 1975 pledge of then–Du Pont chairman, Irving Shapiro: "Should reputable evidence show that some fluorocarbons cause a health hazard through depletion of the ozone layer, we are prepared to stop production of the offending compounds."

"It is time," the senators said bluntly, "to fulfill that pledge." Du Pont should cease production of CFC's immediately.

On March 4 they received an incendiary reply from Du Pont over Heckert's signature:

> Du Pont stands by its 1975 commitment to stop production of fully halogenated chlorofluorocar-

bons if their use poses a threat to health ... [but] ... At the moment scientific evidence does not point to the need for dramatic CFC reductions. There is no available measure of the contribution of CFC's to any observed ozone change. In fact, recent observations show a decrease in the amount of ultraviolet radiation from the sun reaching the United States.

The words were infuriating, perplexing, and in the words of one staff member, "inherently dishonest." The so-called study referred to came from a February 12, 1988, paper in *Science* by the National Cancer Institute, who had simply reviewed the readings from a nonstandard, non-Dobson collection of UV-B meters arrayed around eight American cities and concluded that ultraviolet light had decreased between 1974 and 1985, without considering the effects of increasing smog and the UV-blocking potential of all forms of increased urban pollution.[6] The authors of the UV-B report had taken pains to deny in advance that their results lessened the ozone threat from CFC's, yet their work was promptly misused to prove exactly that. Despite the fact that the paper was effectively useless in the serious debate over CFC's and ozone loss potential, and despite the reality that it had been dismissed by the majority of the serious atmospheric scientific community, the Du Pont corporation (which fancied itself a bastion of responsibility) proceeded to gleefully use the report to counter the avalanche of evidence of the past twelve months!

"Incredible!" roared one environmentalist in particular, who, by his own admission, was becoming increasingly cynical by the day. "What the hell does it *take*? Apparently they're determined to keep selling CFC's until they have irrefutable evidence that CFC's have, in fact, killed all forms of life on the planet. Then and only then will they stop production. Anything short of that would obviously be considered inconclusive. Incredible!"

Du Pont had set itself up for an embarrassing fall. Within the month their chairman's letter would be hung around their collective necks like the deceased albatross of the ancient mariner, badly tainting later public relations attempts by Du Pont to paint an image of itself as having struggled mightily to stay up with the latest science and act immediately in the public interest. In the meantime, scientists such as Du Pont's own Mack McFarland—who knew the score—could only cringe in silence.

On March 14, 1988, the U.S. Senate ratified the Montreal Protocol by a steamroller vote of 83 to 0. The Protocol might not go far enough, but even to knowledgeable senators such as Stafford, Baucus, Durenberger, Gore, and Bumpers, who were determined that the United States do far more, it was an international start, and it was an agreement that had a built-in process for strengthening its provisions as the science became more overwhelmingly certain. Voting against doing *something*, in the distant hope that a tougher international pact could someday be concluded, was unrealistic and unthinkable, and even the environmentally knowledgeable senators supported the bird-in-the-hand pact, which at least acknowledged that "Chicken Little may have a point."

The name of Chicken Little had been invoked so many times by chemical industry representatives scornful of warnings about possible dire effects of CFC's on the Earth's life-supporting atmosphere that even journalists were beginning to forget the fictional little character came from a children's proverb, and not some university research lab. "There they go again!" had been the standard eye-rolling response to any new refinement of the CFC–ozone destruction theory and its consequences, especially if enunciated by Sherry Rowland.

On March 15, 1988, however, Chicken Little's sky did indeed fall—on those in the chemical industry who still held out any substantive hope that CFC–caused ozone destruction could be kept to acceptable levels with continued production of the chemicals. The crushing weight of the Ozone Trends Report was too much.

The results had become so startling over the previous months that the Ozone Trends Panel members had suddenly become alarmed after a November meeting and redoubled their efforts to be absolutely sure they were seeing what they thought they were seeing in the data. "Before announcing something like *that*," said Watson, "we had to make sure we were absolutely right."

They were, and now it was time to tell the world.

In a nutshell, Don Heath's TOMS data had told only part of the story. It was well understood that the satellite-based spectrometers aboard Nimbus-7 had degraded over the past decade, but they still gave a sufficiently stable baseline of readings about atmospheric ozone from which the Panel could judge the ground-based Dobson network readings, culling out the bad and the unreliable readings that didn't show up in the satellite readings. By the same token, once the faulty Dobson stations were taken out of the loop, the good ground-based Dobson readings were used to calibrate the satellite readings. In effect, the systems bootstrapped each other, and in the end the Panel found that Don Heath's conclusions about how much global ozone had been lost were conservative: The overall loss in the Earth's ozone layer wasn't 4 percent as the TOMS data had shown; it ranged as high as 6 percent!

Bob Watson's consummate skill at political balance showed once again in the timing of every aspect of the release of what constituted the executive summary of the report. On the evening of the day the Senate had unanimously ratified the Montreal Accord, Watson began the process of alerting the Reagan White House of what was coming. There was no way in the face of an 83 to 0 vote that the Reaganites could block the Trends Report or dull its effect even if they tried, but with Watson's timing, there would be no opportunity to do so either.

Yet it would be improper (and politically dangerous) to blindside the administration, so the next morning—only two hours before their press conference—Watson personally briefed several senior EPA people along with the presidential science adviser, Bill Graham (who rightly or wrongly had acquired a reputation for a "go slow" philosophy toward CFC controls).[7] That took care of both the official scientific representative and the scientific policy-making sectors of government. At the same time—with one hour to go before the briefing—Mack McFarland briefed Du Pont and the Chemical Manufacturers Association on what was about to hit the fan. Not surprisingly, the top leadership of the company missed the significance of it all, if they even considered it to begin with.

The Trends Panel had used the Rowland-Harris method of breaking down the Dobson data, but had done so on a far larger scale. They had reexamined the Dobson network and broken down the trends in terms of seasons and latitude bands. When the Dobson network data had been averaged—because of the method used in averaging—there was no loss.[8] When it was pulled apart, massive ozone losses jumped off the page for various latitudes at various times of the year, and especially for the winter. The losses in the northern hemisphere were twice as large as the models had predicted, and even larger than Sherry Rowland had originally warned. In fact, the level of global ozone loss was even greater than what was supposed to occur under the Montreal Protocol limitations *by the year 2075*—and this was 1988! In the latitude bands that cover the United States, Canada, Europe, China, Japan, and the U.S.S.R., very significant and worrisome losses had occurred each winter season from 2.3 to 6.2 percent and annually from 1.7 to 3 percent. The figures were far too great to be confused with statistical "noise." These were genuine. The residents of the northern hemisphere—billions of humans—were already receiving increasing doses of UV-B radiation due to one single factor: the insertion into the stratosphere of chlorine, borne by man-made chlorofluorocarbon molecules.

Further, the wintertime loss could indicate that cold weather heterogeneous reactions similar to those in Antarctica (but on a smaller reactive scale) could be occurring over mid-latitudes![9]

In addition, the Trends Panel pointed out that the approaching peak of the solar cycle (1991) would tend to increase the atmospheric production of ozone and mask the effects of chlorine destruction of ozone, but after 1991 the effects would begin to become even more apparent, the Antarctic and Arctic ozone holes would grow deeper and more pervasive, and the worldwide ozone losses could be expected to accelerate. If the CFC growth rate continued, the report added, the Earth's population could expect 10 percent of the ozone layer to be gone by 2060.

The effect on Du Pont was much more rapid and decisive than anyone had expected. After fourteen years of unyielding resistance to a challenge that could affect at best 2 percent of their business, Du Pont suddenly did an about-face, leaving the members of their CFC division stunned.

Mack McFarland had begun the process of getting Du Pont's leadership to understand the full significance of the findings after the press conference on March 15. At first, the senior leaders of the corporation had not understood. On March 18, however, McFarland got the chance to brief Chairman Heckert and the corporation's executive committee in person. The meeting ended with a simple decision: Du Pont's production of CFC's would end *as soon as* substitutes became available. With that decision, the corporation's considerable public relations machine began putting a positive-image spin on the upcoming announcement. Du Pont would embrace the decision as if it sprang from the heart of an overly concerned and conservative guardian of the environment.[10]

The fact that it had taken so many years of foot-dragging to get to that point would remain a matter of external debate, but in the end the decision would become a windfall of new profits for Du Pont. Although the company had long been responsible for supplying 25 percent of the planet's CFC's, and 25 percent of the chlorine atoms involved in ozone destruction, now it could introduce new "environmentally safe" CFC substitutes at considerably higher profit levels than could ever be achieved with standard CFC's. Du Pont's leaders could see that the race to replace CFC's would eventually become frantic, as international regulations chased such products from the marketplace. The stampede of industry to switch to substitutes such as HCFC's (which had already entered the testing stage) would make the new replacement gases supply-sensitive, not price-sensitive. Du Pont, in other words, would be ready, willing, and able to cash in. The Earth's ozone layer and its human population might be losers in the long ozone war, but the Du Pont corporation would not be counted among those casualties.[11]

By the time the Snowmass conference finally convened less than two months later, the remainder of the CFC industry was rushing to examine Du Pont's action, the Trends Report findings, McFarland's interpretation, and the results of the Antarctic Airborne Expedition itself in order to decide whether to follow suit. To say the CFC industry was in some disarray would probably be an understatement.

For that matter, though, the Snowmass conference was somewhat in disarray as well.

Sixty-nine presentations—most of them backed by papers that would be refined and published in a final compilation many months later—were scheduled for a five-day period, and each group representing each paper had selected one of its number to take the microphone and explain their findings and conclusions. Those presentations were supposed to be strictly limited to a given time period (usually twenty minutes), but with the passionate interest and attention levels high (and with certain scientists spring-loaded to argue, question, and discuss everything from overall conclusions to niggling details), the various "presiding" scientists (the designated moderator was different for each half day) quickly lost control of the schedule.[12] While the dynamicists staged no revolts and never varied from gentlemanly discussion, neither were they about to give up, and scientists such as Ka Kit Tung of Clarkson University (who had been responsible for the major thrust of the dynamical theory), Jerry Mahlman, Mark Schoeberl of NASA, Cambridge's Mike McIntyre, NASA's Richard Stolarski, and others worked hard to make certain the chemically based explanations presented on the third day weren't allowed to run away with the conclusions. For that matter, when the dynamicists had their turn on Thursday (day four), the chemists returned the favor, Susan Solomon (for one) neatly and skillfully disassembling one dynamicist in particular who had made the mistake of letting his evidentiary guard down while trying to reelevate the dynamics of the Antarctic ozone hole to unrealistic heights of importance.

Of course, behind the scenes was where the important exchanges took place—hallway conversations; evening get-togethers in many of the ski lodge, condominium-style rooms occupied by the scientists; discussions held while jogging, having dinner, playing golf, or simply leaning against a pickup truck in the parking lot during the lunch break. The formal presentations and papers and posters during such a conference form the basic stock for refining a major scientific issue, but the person-to-person exchanges provide the fermentation from which ultimate consensus—and new realizations—emerge.

On the fifth day, with everyone's ears full of words and heads swimming with ideas, there had in fact been general movement of the group in several directions. It was obvious, for instance, that while dynamics played a key role in Antarctica and could not be dismissed or minimized, chlorine chemistry was the key. Tuck's cornfield meet had indeed occurred.[13] There had been little consensus, but from the clashes had emerged a consistent understanding.

Moreover, there was no question that the urgency of the CFC situation was acute, whether or not the Antarctic hole could spread to lower, more populated latitudes. The fact that anything had gone nonlinear in the formerly balanced equation of atmospheric ozone creation and destruction was now established fact, as was the role of heterogeneous atmospheric reactions (thanks to the pivotal laboratory work in 1987 by Mario Molina). And from all that came many new understandings, including the fact that the Arctic Airborne Stratospheric Expedition scheduled for 1989 was indeed vital. If there was a hole in the ozone layer each winter over the northern ice cap, however weak it might be, it could already be affecting millions of people.

For Susan Solomon there were few real surprises at Snowmass. Being on the leading edge of her subject meant her role was one of refinement and support for the various theoretical explanations that were now jelling nicely.

There was, however, one rather profound shock—the hint of a dark possibility—that arose in her mind from the words and the presentation of a scientist from Menlo Park, California.

Dr. Margaret A. "Maggie" Tolbert, a bright young scientist from SRI International in Menlo Park, California, had presented a paper on the second day of the conference about heterogeneous chemistry in the stratosphere related to Antarctic Ozone depletion. Her early work had concerned the tiny chemical reactions on the face of the ice crystals in the PSC's. Her new studies showed that there was no guarantee the very same reactions at the very same speeds couldn't occur on other tiny surfaces.[14]

Certainly there was no real possibility of Polar Stratospheric Clouds over mid-latitudes of the Earth. Susan's January trip to Thule, Greenland, had given her reason to believe that PSC's were generally rather minimal even over the north polar regions. So to the ear of the casual observer at Snowmass, Tolbert's work on heterogeneous reactions could not really concern the bulk of the Earth's human population.

Or could it?

Susan Solomon had already read extensively about the various mixture of particles that could from time to time be injected into the stratosphere. Most of them were in volumes too small to worry about, but there had been much speculation that the same type of reactions that took off at a high rate of speed on the surface of the PSC ice crystals might find a similar reaction crucible on the surface of volcanic debris.

The Mexican volcano of El Chichón in 1983 had injected large amounts of tiny particles in the stratosphere that might have caused (through heterogeneous reactions) a small ozone drop.[15] El Chichón was a big eruption, but there had been other historical eruptions far larger, such as the explosion of Krakatoa in the nineteenth century. That's where a cold feeling of genuine worry began to crawl around the back of Susan's mind, expressed in a single question she began asking herself: "What if we have another Krakatoa?"

"I wasn't scared," Solomon would explain later, "until I saw Tolbert's paper. People like Tolbert who have done these surface reactions are now starting to do them on sulfuric acid aerosols at warmer temperatures, which is what you have, for instance, over Boulder. And you know what? They're finding that the reactions go [at greatly accelerated rates], and that's really frightening." The sulfuric aerosols Tolbert had considered took the place of PSC's in providing a reaction surface, though exactly what the potential was would have to wait for more research.[16] Susan had begun thinking by the end of Snowmass of what she could do to push the research further, including, perhaps, collaborating on a paper, perhaps with Dave Hofmann, who had studied these perhaps for more than a decade. If there was any possibility that such nonlinear reactions could really occur in moderate temperatures over mid-latitudes, the sooner the information was confirmed and given to the policymakers the better.

"I mean," she said, "if it's only going to be a polar phenomenon, clearly that's good in a lot of ways . . . but what if you *do* have Krakatoa? We've loaded the stratosphere with chlorine, with more arriving every hour and for the next fifty years at least. The world could literally be a time bomb. The bottom line is that we've been putting chlorine into the atmosphere on a time scale that is minuscule compared with the geologic time scales. With the current loading of chlorine in the atmosphere, and if these people are right about those surface reactions taking place on those aerosols, you could see some changes here that have the potential for biological catastrophe. I'm not saying that if something like [Krakatoa] were to take place it would wipe out life on Earth as we know it, but it would have serious crop damage effects and could easily cause other kinds of agricultural disasters. Obviously right now [these theories] are very uncertain and we need to understand the chemistry of these aerosols a lot better than we do, but it's frightening. And *I'm* frightened."

The last word at Snowmass, figuratively and literally, went to the battered master who had quietly and calmly stayed his course for fourteen years, Sherry Rowland. Snowmass had in some respects provided the final validation of the concerns raised by Sherry Rowland and Mario Molina's work, expanding exponentially the original finding that CFC–borne chlorine can destroy ozone.[17] In terms of the scope and breadth of the atmospheric chlorine chemistry presented by scores of scientists within the five-day conference, the meeting was pivotal—but it was still Mario Molina (who spoke at the start of the last session on his latest chlorine species research) and Sherry Rowland who had originally sounded the alarm. Of the two, Rowland had been the lightning rod.

With his half glasses and sonorous voice, Rowland's presence as always was soothing, his image that of the steady, unflappable professor. He rose to the dais as the last speaker of the Polar Ozone Workshop and threw down a new gauntlet before his fellow scientists (some of whom were former students). While the chlorine-ozone issue was now thoroughly exposed (if not yet resolved) there were other major challenges on the stratospheric agenda that Sherry Rowland knew the atmospheric scientific community had to face, and the startling and somewhat inexplicable rise of methane levels was one of them. Methane, which comes from the guts of cows, termites, and rice paddies among other anaerobic (without oxygen) sources, is a greenhouse gas when it stays in the troposphere. Exactly why it's increasing no one really knows, nor is there a definitive explanation for what source is causing the worldwide measurable increase. But with more methane in the troposphere, global warming speeds up.

Sherry Rowland had identified another worry. Methane (CH_4) in the troposphere ends up transported upward through the tropical tropopause (near the Equator) into the stratosphere, and, once there, breaks down to form water (H_2O), which in turn can increase the worldwide concentration and duration of polar stratospheric clouds, which provide more of a reactive surface for ozone destruction.

Rowland (who had worked on the problem with one of his graduate student research associates, Donald Blake) delivered a short paper on the subject and sat down, saying not a word about the main focus of the conference. The fight was not over. CFC's were still being manufactured and the world was still spewing carbon dioxide skyward at a furious rate while methane and nitrous oxide concentrations increased daily. Snowmass marked a milepost, not a destination.

In other words, the battle for the health of the Earth's atmosphere has barely begun, and because of the economic and political forces that will resist change in the absence of actual evidence of damage, without a vibrant and engaged atmospheric scientific community energetically looking for the evidence and the answers, the ozone debacle will be repeated. Nothing will be done until major, possibly catastrophic, damage has already become inevitable.

Within a month, another quiet and studious scientist of world-class reputation would sit before a sympathetic Senate subcommittee and offer a reasoned opinion that would put his career, his reputation, and his credibility under worldwide attack, all for expressing a singular thesis: The effects of global warming are already with us.

As the Snowmass delegates departed—and the temperatures around the United States climbed alarmingly—the stage was once again set by nature for a play with all the same plot features of the ozone trilogy. There would be theories and scathing, sneering rejection of the theories and the theorists; there would be environmental groups pitted against industrialists and politicians; there would be an old president and a new president both moved to dynamic inaction in the face of conflicting scientific evidence. Once again there would be a press, a public, and a political establishment unable to understand what science was trying to communicate.

But this time, the global and national stakes would be even greater.[18]

CHAPTER

9

Strangers in a Strange Land

Dirksen Senate Office Building, Washington, D.C., June 23, 1988

It was familiar territory, and as expected, there were reporters waiting to interview the director of NASA's Goddard Institute for Space Studies as Jim Hansen approached the door to the hearing room. Hansen had turned down interview requests from radio and television the day before, passing on the opportunity to do ABC's *Nightline* with Ted Koppel to co-worker David Rind. Rind and Tony DelGenio, both GISS scientists, were more comfortable talking to the media. In fact, Jim Hansen more or less dreaded doing television, though he knew it was necessary from time to time. After all, what he had come to Washington to say on the record had already leaked to the media in written form, and even the careful, technical, administration-approved language couldn't completely mask the fact that if the GISS models were telling the truth, global warming was about to become a far hotter issue than ever before.

Jim balanced his dilapidated shoulder bag (which had become de facto briefcase, overnight bag, and a bit of a trademark) and greeted the newsmen quietly, answering their questions one by one until a member of Senator Tim Wirth's staff materialized at his side.

"Dr. Hansen?"

"Yes."

"Senator Wirth would like to speak with you for a minute before we get started. Could you follow me, sir?"

Jim Hansen nodded at the young man and excused himself from the reporters, not at all unhappy to leave the questioning.

His mind was on the hearing to come, and the questions had been a distraction. His formal, prepared statement for the record had long since been finished, of course. That had been written weeks before in his New York office and properly submitted through NASA channels for administration review. But what he would say *verbally* in the hearing room would require a brief summary of the more formal, prepared statement, and that summary had to be carefully phrased. Jim Hansen didn't like to ad-lib, especially on such an important subject.

Senator Tim Wirth from Colorado was waiting for him when the aide ushered Hansen through the door and disappeared. Wirth, an athletic man with a mop of sandy blond hair and a broad smile, looked younger than his forty-nine years. He had read Hansen's prepared statement, he told the NASA scientist-administrator, and wondered if Hansen would mind if Wirth reshuffled the order of witnesses to put him up first.

"Sure," replied Hansen. The request wasn't terribly surprising, in view of the media interest, but it gave him a momentary case of butterflies, the normal jitters any player feels before taking the field for a pivotal game.

And this might just be a pivotal hearing.

Jim Hansen regained the hallway and walked back toward the hearing room in thought, recalling the suggestion he had made to a staff member of the Energy and Natural Resources Committee before and after a November 1987 hearing on the same subject. "If you want a congressional inquiry on global warming to have real impact in the media," Hansen had told him, "hold it during the summer when it's hot. It's hard for people to relate to this issue in winter." They had listened well. Washington was broiling in a record heat wave, and Jim knew he had a rare opportunity at hand: With the GISS three-dimensional climate model of the greenhouse effect finished and the paper published at last, it was time to make a stronger statement than he had been able to make in 1986 before John Chafee's committee or in 1987 before another Senate hearing. Senator Wirth, it seemed, had provided him a perfect stage.

In fact, Wirth and his committee staff were either very lucky or clairvoyant. Many months before Washington temperatures began to get warm, they'd decided to schedule this hearing on the anniversary of the hottest day on record in Washington, D.C.'s history. But *this* June 23 was shaping up as one of the hottest days of the hottest summer of the present decade. The Midwest was locked in the jaws of a horrendous drought, the Mississippi River had already dropped so low barge traffic was coming to a halt, crops were dying in the fields, and thermometers in the South and Southwest were pushing over the century mark day after day. The public was thoroughly alarmed, with memories of the 1930s dust bowl years amply replayed on the evening television news broadcasts that had begun blaring the hottest temperature and crop damage as a nightly item. Hot weather had always had the ability to bring a sweltering U.S. population from a state of consternation to the edge of exasperation during any particularly rough summer, but this was an exceptionally hot and arid one, and it was only June. Plus, when sharply focused in the impassioned and bloodshot eye of network television, the trail of record temperatures led to a possible villain for the first time in anyone's memory—a common enemy that this time just might take the blame for all the heat and misery: global warming. The mission of the electronic and print media was clear: explore the hypothesis that the heat of the summer of 1988 was caused by the greenhouse effect. With those marching orders, a Senate committee hearing on "Greenhouse Effect and Global Climate Change" was made-to-order for excellent sound bites and lead stories, if the witnesses or senators said anything even remotely quotable. That didn't mean the media was being devious or planning to invent a story. It merely meant the reporters and their producers and editors were going to hang on every word spoken by anyone in a position of scientific or official knowledge in hopes of providing *the answer* to the American electorate.

Jim Hansen, of course, was both scientific and official—an official of NASA, no less—and Senator Wirth and his staff were hardly babes in the woods when it came to arriving at such realizations. The lineup of witnesses included representatives of two responsible environmental research groups, the director of a private research center, a scientist from NOAA, and Hansen, the director of the NASA Goddard Institute for Space Studies—a man who was already on record as believing global warming was under way as a result of the greenhouse effect, and one whose prepared statement contained some juicy quotes. Even a freshman media student could instantly identify Jim Hansen as the logical lead witness for maximum effective media coverage.

Hansen didn't live in an ivory tower, but in some respects he tried to work in one, quietly pursuing advanced atmospheric computer modeling and planetary atmospheric studies from within the slightly worn walls of the Goddard Institute and his corner office overlooking the uptown New

York corner of 112th and Broadway. The world was out there, all right, but his desk—and his attention—faced inward. He had never felt comfortable in the glare of publicity, and was considered naïve in the ways of the political world. But he was too intelligent and accomplished to be unaware of the world realities, so at 2:10 P.M., as he walked into the hearing room and noticed reporters and TV cameras, there was no question in his mind that the media was responding to the common question of overheated Americans: What the hell is going on with the climate?

What he did not realize, however, was the media's depth of determination to provide a simple, unambiguous answer, appropriate for a sound bite of TV coverage or an attention-grabbing newspaper headline that would cut through the type of language newsmen considered scientific doublespeak and make clear sense.

Hansen sat down and pulled his papers out of his hang-up bag including his handwritten notes and his prepared statement for filing in the record. It was amazing, but this time OMB hadn't slashed his formal presentation at all, and that was certainly a refreshing, unprecedented, and odd change.

The memory of his encounters with the faceless, nameless OMB censors was all too fresh. In June 1986 when he had testified along with Sherry Rowland and Bob Watson on the greenhouse effect—when he had laughingly apologized for sounding like a "befuddled scientist"—there were several recommendations he had really wanted to make. Chief among them was his determination to sound an alert about the desperate need for the nation to change its scientific educational policy and start throwing heavy financial support behind the development of young scientists from undergraduate through the postdoctoral years, in order to prime the pump of brainpower to solve some of the perplexing problems they were facing in atmospheric science and other areas of science. A Senate panel was a good place, he figured, to recommend that the United States go back to the type of dedicated support of young scientists that had been around in the sixties when NASA was priming the educational pump with serious financial commitment. That commitment was all but gone now, and Jim Hansen, as a NASA unit administrator with responsibilities for keeping good scientists working in the Goddard Institute, had been struck by how thin the supply of top-quality young postdocs was becoming. There were thousands of sharp minds out there, but they were following the money into law and medicine and engineering rather than taking a vow of poverty as a scientist. Jim Hansen wanted to change that, but his formal statement came back from OMB and NASA headquarters with the recommendations slashed out. "Not in accord with administration budgetary policy" was the explanation. His recommendations would mean more federal money for science during the term of a president publicly dedicated to slashing federal spending in all quarters. It was financial heresy, as OMB's deletions had confirmed.

"But," said a friend at NASA headquarters, "you can say the same things verbally, if you have time, and if you qualify it as only your opinion as a private citizen."

He had done so, but was dissatisfied with the result, and although Chafee's "King" question had presented a golden opportunity, Hansen had failed to seize the moment and respond with a clear, strong statement about the potential dawning of measurable global warming.

In 1987 another congressional hearing invitation landed on his desk, and another formal statement was written and forwarded to D.C. This time it came back looking as if it had detoured through a shredder. The OMB censors had wielded the sword of Reaganite ideological purity with such effectiveness, they had neatly emasculated virtually every significant point Hansen had tried to make regarding global warming and climactic change potential. The pablum he got back made it look as if the NASA Goddard Institute for Space Studies knew absolutely nothing about the greenhouse effect, despite tens of millions of dollars and many years of cutting-edge research involving not just Earth's atmosphere, but the atmospheres of Mars and Venus as well. In utter disgust, Hansen had asked for permission to throw away the ruined statement and testify as a private citizen, and for some reason, OMB agreed with the request. Considering the changes they had attempted to make, OMB couldn't have liked what Hansen presented. Here he was, a government scientist, arguing forcefully that a "wait and see" greenhouse policy was extremely dangerous because it builds in future climate changes without understanding the consequences. But again Hansen was dissatisfied with the hearing, which received little attention and had no measurable impact.

And then the invitation for the June 1988 hearing had come in. At last they were going to hold one in the heat of summer. Jim Hansen had written what he felt was a carefully metered statement and sent it in to Washington, expecting the same sort of tattered, red-streaked, overedited remains to come back.

Instead, the statement disappeared into the maw of bureaucratic oversight only to boomerang back in near-virginal condition. Perhaps they had made a painful decision, Hansen concluded, that another slashed statement would unleash private citizen Hansen once again to sound off before the Senate, the press, God, and the whole world about matters the administration felt were best left to less passionate researchers. It was safer, it seemed, to have Hansen on the record and under *some* control than let him speak "privately" in public. No matter how much a NASA scientist prefaced his remarks with warnings that he was speaking as an individual, he was still a leading NASA scientist, and everyone knew it. It was not an unreasonable assumption that such a man would not reach a conclusion, personal or otherwise, without basing it at least in part on his official research.

Passionate was a word that didn't seem to fit Hansen. Dedicated, professional, intensely honest, yes. Passionate in its classic sense, no. But when it came to his alarm at the drying pipeline of scientists, passionate probably did apply.

Hansen himself had been the beneficiary of those sixties programs. A native Iowan, he had graduated from the University of Iowa in 1963 with the highest distinction in physics and math and had been accepted as a NASA trainee. For the next three years, under the wing of NASA and with great enthusiasm and excitement, he sailed through his master's degree program in astronomy (again at the University of Iowa), spent a year as a visiting student at the Institute of Astrophysics at the University of Kyoto, Japan, and received his Ph.D. in physics (University of Iowa, 1967). Thus accredited at the hands of a government program, he was ready to conquer the world—or at least the atmosphere of a neighboring world—in this case, Venus.

GISS was a quiet satellite offshoot of the large NASA Goddard facility in Greenbelt, Maryland. Largely ignored in the grander scheme of NASA operations and glamour, tenuously funded, and out of the mainstream of the manned space program, GISS had nevertheless become a theoretical center for atmospheric research on unmanned planetary probes, and to join the GISS team—as Jim Hansen did in 1967 as a postdoc resident research associate—was both ideal and exhilarating.[1]

While Hansen worked on the *Pioneer* spacecraft, an atmospheric scientist named V. Ramanthan discovered that CFC's too, were greenhouse gases, and far more effective ones than CO_2. Hansen read Ramanthan's paper with great interest. CFC's were not only suspects in ozone depletion, they were thousands of times more effective in trapping the Earth's infrared heat than carbon dioxide. When a young postdoc named Yuk Ling Yung of Harvard asked his help on looking for other greenhouse trace gases in 1975, Hansen agreed, publishing a joint paper the next year that showed that methane, nitrous oxide, and several other gases also soaked up infrared radiation bouncing back toward space from the surface of the Earth. The work made it clear that the greenhouse effect was going to be even more important on Earth than people had previously suspected, and the effects of increasing CO_2 concentrations should be obvious sooner.

With the theoretical foundation laid, Jim Hansen's interests turned to the Earth's atmosphere exclusively, and he began focusing on the greenhouse effect and the fact that anthropogenic greenhouse emissions of CO_2 from the burning of fossil fuels since the Industrial Revolution had raised the concentration of CO_2 in the atmosphere from under 280 parts per million to over 350 parts per million (about a 25 percent increase). Much of his previous research had involved tedious calculations of how forms of radiant energy from the sun were scattered and absorbed by the gases and particles in planetary atmospheres. Now, though, a new world of advanced, high-speed computer modeling of climate conditions had opened up, and he resigned as the principal investigator of the Venus probe and began the drive to develop a three-dimensional mathematical model of the Earth's climate.

The idea was simple, but the execution was incredibly complex. Using the fastest computers available with immense number-crunching power, they would program in sophisticated sets of formulas that were supposed to accurately predict the atmospheric processes of heat and energy transfer that drive weather in the troposphere (the lower atmosphere). If they could refine the formulas and get them right eventually, the weather conditions for any given day for the entire Earth could theoretically be programmed into such a model, and when the modeling computer program began, it would methodically print out the resulting conditions for each successive day in the future, giving temperature, storm, and rainfall patterns. The exact weather pattern for any specific day in the far future (in other words, more than a week or two in computer-modeling terms) is in fact unpredictable because the atmosphere and ocean are chaotic fluids which slosh around. But the climate, the *average* weather and its statistics of variability, *is* potentially predictable *if* the model becomes sufficiently sophisticated and realistic. Weather prediction models had been around for fifteen years, but Hansen's team at GISS wanted a model that could simulate the climate over decades as CO_2 concentrations increased slowly and steadily. To simulate such long periods with the GISS computer, each cell or sector of the resulting model had to be very large in area to keep from exceeding the available computer capacity, which meant, for instance, that all conditions in a space the size of a large state would be represented by average values for temperature, cloud, and moisture conditions. Even when clouds *could* be handled in the model, they had to be considered flat and huge. Obviously the real atmosphere doesn't work that way, so the resulting pictures of what the world's climate might be like in five, ten, twenty or more years were highly suspect, especially when such monkey wrenches as doubled CO_2 or added CFC concentrations were programmed into the calculations.

It wasn't environmental concern that drove most of the climate-modeling scientists, it was the scientific chase—the opportunity to look at a planet with an atmosphere in the throes of massive chemical and gaseous change for the purpose of determining what was going to happen, and what, if anything, could be done.

At first, NASA funded the development of a three-dimensional climate model at GISS and the application of it to the greenhouse problem. But in 1980 the Department of Energy (DOE) was named as the lead agency for CO_2 research in the United States, and NASA had to divert research money to pay for a supercomputer at the Goddard Spaceflight Center in Maryland. Funding for the CO_2 climate studies at GISS suddenly was cut off, with the recommendation that Hansen seek support from the DOE. Fortunately, DOE agreed to pick up that task. Unfortunately, after President Reagan took office in January 1981, new appointees took over the climate program at DOE, and those new appointees brought with them a distinct distaste for environmental "alarmism" (as one Reagan staffer broad-brushed much of atmospheric science). It was time for those running research projects that could be considered politically controversial to lie low, but Hansen and his researchers, who were not adept at reading political tea leaves, made a critical "mistake": They committed what he would later refer to as his "original sin" by publishing a paper in *Science* that simultaneously made the front page of *The New York Times,* and instantly angered the administration.

According to Hansen's paper, the world wasn't getting colder (there had been a cooling trend between 1940 and 1965 that had led to some predictions of an impending ice age), it was getting warmer, and the warming trend they had identified was roughly consistent with the global warming predicted by the greenhouse effect. That might not have been so bad, but because their paper, and *The New York Times* article, described the possibility of longer growing seasons, the possible opening of the "fabled northwest assage" due to shrinkage in the Arctic ice cap, more frequent droughts, and rising sea levels as a result of CO_2 emissions, Hansen—and GISS— became by definition environmental alarmists.

The reaction was swift. The DOE managers criticized the quality of the Hansen paper and the GISS global climate model in general, apparently setting the stage to justify their next action: There would be no DOE funding at all for Hansen's climate modeling. In addition, they decided to phase down their entire CO_2 research program from 12 million to 8 million dollars.

But that was too big a move to escape notice on Capitol Hill. A young representative from Tennessee, Albert Gore, became aware of DOE's plan and ordered a House hearing on the greenhouse effect and the DOE program. Hansen and several other scientists journeyed to Washington for the hearing, which gave Gore the momentum to force DOE to change its decision.

There was still no chance, however, that DOE was going to supply any funds for James Hansen. In fact, it appeared that they pointedly favored researchers who could pass the litmus test of disputing Hansen's 1981 paper. In other words, those who believed there was such a thing as global warming need not apply at DOE.

In any event, Hansen, who had been appointed director of GISS in early 1981, now had to begin laying off his scientists and postdoctoral assistants and, in effect, putting the project on the sidelines while he searched for new support from somewhere in the environmentally hostile administration, a process that took nearly two years. Finally, a renegade band within the Environmental Protection Agency—renegade because they felt strongly enough about the issue to stick their necks out and quietly re-fund the GISS project—put Hansen and the GISS CO_2 research back in business.

Later in 1983, however, the EPA—with GISS assistance—went too far in the eyes of the Reagan administration. They issued a report to the nation on the greenhouse effect that seemed to suggest the need for actions that might in some ways impact the business community, and the report predictably made more headlines, and more enemies at the White House. Even the science adviser to the president denounced the report as "alarmist" (which was becoming a favorite Reaganite term to describe any environmental statement or finding), and instead, the White House embraced an alternate report by the comparatively tame National Research Council that adeptly reviewed the science, then contradicted its own findings by glibly assuring everyone that mankind had at least twenty years to mull over the greenhouse–global-warming problem before action might be necessary.

GISS was sent back to the Reagan doghouse in political disgrace and required once again to search for funds. The political shunning gave Hansen every reason to become even quieter and more cautious than his fellow scientists already knew him to be. But he still felt compelled to state the realities as he saw them when asked, and saw no reason to change. Politics wasn't his game, science was, and in his opinion, even being director of GISS shouldn't require him to dissemble or sidestep questions about their findings, when ultimately it was the public who paid Jim Hansen's salary. In the 1986 hearing with Bob Watson and Sherry Rowland, he had guarded his statements very carefully—in part because some of their key research papers were not complete and because he personally was not strongly convinced that a "signal" had emerged from the noise of natural climate variability. In 1987 he had been more sure of what he was seeing and what should be said, and his testimony was somewhat less guarded, but still not incendiary. In fact, before 1988, the statements of Dr. Jim Hansen, director of the Goddard Institute, had gone relatively unnoticed by the press and the public.

But as Senator J. Bennett Johnston of Louisiana opened the hearing that Tim Wirth would chair on June 23, 1988 (and as the thermometer just outside the Dirksen building passed 94 degrees and was still climbing), all that was set to change.

"We have only one planet," said Senator Johnston during his opening remarks. "If we screw it up, we have no place else to go."

Wirth took over at that point and waded in with his own opening:

> Meteorologists are already recording this as the worst drought we have experienced since the Dust Bowl days of the 1930's. The most productive soils and some of the mightiest rivers on Earth are literally drying up... Already more than 50 percent of the northern plains' wheat, barley and oats have been destroyed, and the situation could get much worse. On Tuesday, the Mississippi River sank to its lowest point since at least 1872 when the U.S. Navy first began measurements. And in my home State of Colorado, peak flows are among the lowest on record, and reservoir levels are also alarmingly low.

We must begin to ask is this a harbinger of things to come? Is this the first greenhouse stamp to leave its impression on our fragile environment? I understand that Dr. Hansen will provide testimony this afternoon that points clearly in that direction.

Indeed he would. With the television cameras silently recording and journalists' pens poised, Jim Hansen calmly strolled into the maelstrom.

I would like to draw three main conclusions:
Number one, the Earth is warmer in 1988 than at any time in the history of instrumental measurements.
Number two, the global warming is now large enough that we can ascribe with a high degree of confidence a cause and effect relationship to the greenhouse effect.
And number three, our computer climate simulations indicate that the greenhouse effect is already large enough to begin to affect the probability of extreme events such as summer heat waves.

Good, but not punchy enough. The journalists strained forward, waiting for key words, catchphrases, and sound bites as Hansen began explaining point number two, that finding a causal association between the greenhouse effect and global warming required the global warming to be greater than "natural climate variability" (background noise, in statistical terms), and that the warming had to be consistent with the predicted operations of the greenhouse effect.

Another slide went up on the screen as Hansen carefully traced the computer model predictions of what might happen with increased CO_2 in the atmosphere, the incidence of yellow and red indicating hotter-than-normal summers beginning to dominate the picture of the years 2025 and beyond. *Had global warming been statistically validated?* Well, said Hansen, when the worldwide temperature records for the last century are examined, the temperature rise is, on the average 1 degree Fahrenheit (.5 degree Celsius). When you look at the last three decades, he said, there was a significant rise—*definitely* a warming *trend*. "The probability of a chance warming of [this] magnitude," said Hansen, "is about one percent."

Suddenly the media was on alert. The last sentence had twanged something deep in the journalistic early-warning system promising that a taut, juicy, quotable quote was on the way. Within a microsecond half-dozing reportorial minds were straining to hear every syllable Hansen spoke, and he didn't disappoint them:

"So with ninety-nine percent confidence we can state that the warming during this time period is a real warming trend."

Aha! Headlines suggested themselves immediately. "NASA scientist is 99 percent confident that the warming of the Earth is a trend." No, he's 99 percent confident that the warming trend is the *greenhouse* effect in action. Or did he say that he's 99 percent sure the greenhouse effect has caused the present warming, the drought, and the boiling temperature in Washington, D.C.?

Several minutes later, amidst all the proper caveats and limiting statements, Hansen seemed to say it again: *"In my opinion, the greenhouse effect has been detected, and it is changing our climate now."*

Okay! That clarified it. He's sure the greenhouse effect is responsible for changing the climate now, which means today, which means the current heat wave, which means lead story on the evening network newscasts.

Buried at the end of his statement was another important phrase, but one that only a few reporters caught, yet it was the limiting phrase that Jim Hansen innocently believed would prevent the very misinterpretation that had already taken root: "... *we conclude that there is evidence that the greenhouse effect increases the likelihood of heat wave drought situations in the southeast and midwest United States* even though we cannot blame a specific drought on the greenhouse effect [*emphasis added*]."

Even though Hansen said it again during the question and answer period following the last witness ("... you cannot blame a particular drought on the greenhouse effect"), a connection had been made that subtle refinements would not alter.

Phil Shabecoff, one of the better science journalists in the country (and the heir to the throne of *The New York Times'* science writer Walter Sullivan), caught Jim Hansen at the back of the hearing room for a few questions, one of which sparked the response: "I think it's time to stop waffling so much and say that the evidence is pretty strong that the greenhouse effect is here and is affecting our climate now," said Hansen. The line would make *The New York Times'* June 24 edition along with an excellent article that did *not* conclude that Jim Hansen had branded the heat wave and drought of 1988 as "Courtesy of the Greenhouse Effect."

But many of the media were not as careful. The "99 percent" line was simply too good to pass up, and the public implication—especially for fast-news junkies—was that a respected NASA scientist had flatly stated he was 99 percent certain the summer misery was caused by the greenhouse effect, even though Hansen had said nothing of the sort.

The fallout from the scientific community was swift and angry. While the environmentalists were excited and pleased, meteorologists were angered by the sudden preemption of their discipline and upstaging of their quiet research. Others, including many friends and colleagues in the atmospheric science community who respected Hansen, were aghast at his flat statement that a cause-and-effect relationship had been established with a high degree of confidence. Well-known atmospheric scientists such as Dr. Stephen Schneider of the

National Center for Atmospheric Research (NCAR) in Boulder, Colorado—himself a vocal advocate of environmental caution who was often erroneously labeled a radical—felt Hansen had gone too far with the 99 percent figure when he heard of the testimony.

And Bob Watson of NASA cringed. Hansen, he felt, was not reflecting the scientific consensus. They were all ready to concede that global warming was *coming,* and had probably already begun, but, so far, only Jim Hansen had concluded publicly that it was actually detectable and out of the noise level (even though Watson well understood that Hansen had *not* pinned the summer of 1988 on the greenhouse effect). Watson knew instinctively that the 99 percent figure would spark a tumultuous public focus on the subject of global warming, and the focus might be very unfortunate for one reason: It was all tied to the heat of the summer of 1988. With the next cool summer, the alarm would evaporate—along with the political will to address what was a very real problem.

As the grumbling among scientists rapidly escalated into a major thunderstorm of disagreement with what Hansen had said and was supposed to have said, the progenitor of the 99 percent line himself was fighting a losing battle trying to respond to the thundering herd of newsmen who deluged his secretary with literally hundreds of telephoned requests for interviews. Hansen could handle only a small fraction, especially if he expected to get any other work accomplished. He appeared on television newscasts around the country and on the front page of *The New York Times,* and radio hosts nationwide piped his quiet voice through millions of radios in the week that followed, all of it driven by what Hansen was beginning to recognize as a basic misunderstanding of his message. He wasn't saying the greenhouse effect *caused* the heat wave and the drought, but merely that the greenhouse effect was already increasing the *likelihood* of such droughts and heat waves. He was saying the greenhouse effect had caused global warming that had now been seen and identified, but the drought and heat wave of 1988 was not that "identified" proof. He had been relying on the frequency of high temperature years, the overall rise in global temperature averages, and other such subtle tools. But he hadn't succeeded in getting those points across. To the public, the summer of 1988 was a greenhouse–global-warming event identified, charged, indicted, and convicted by the National Aeronautics and Space Administration in the form of Dr. Jim Hansen, a scientist who presumably knew what he was talking about.

To counter the impression and make a better stab at explaining how global warming manifests itself as the increased *chance* of anomalous weather over a period of years, Hansen quickly made up an oversized set of dice with red, blue, and white sides. The red side represented a warmer-than-normal summer, the blue a cooler-than-normal one, and the white an average summer. The first die represented the period up to 1958, and had two sides of each color. But the die representing the near future contained four red sides, one white side, and a green side. The chances of scorching summers would increase markedly when the dice were loaded. But the cause of any one excessively hot summer could be either natural climate variability or the greenhouse effect, he cautioned, and there was no way to tell for certain, any more than one roll of the dice could establish a trend.

Hansen used the dice on several TV interviews, but the explanation fell, if not on deaf ears, then on confused ones. If Hansen was confident he had identified a cause-effect relationship between the increase in greenhouse gases and global warming and had spotted evidence of global warming, then wasn't the heat wave a result of global warming? After all, here was this NASA scientist saying that the greenhouse effect caused global warming, which increased the chances of a hot summer—and 1988 was definitely an unusually hot summer—yet he wanted to hold back from saying that this *particular* hot summer was the direct result of global warming and not a result of natural climatic variation? For most Americans, it was a bit too esoteric a distinction—and that in itself highlighted a national problem.

Even in an ideal world with a perfectly educated nation containing 220 million scientifically literate citizens, human nature would dictate the possibility of at least some public misunderstanding. But if fathoming what Hansen was trying to say required at least a basic scholastic grounding in science and the scientific method, then given the sorry state of science education in the United States, he had surely cast pearls before swine.[2]

For his part, Jim Hansen was surprised at the tremendous outpouring of media attention, and dismayed at the inaccuracy of the reports on what he had said. Some simply misquoted him, most misconstrued his meaning, and a few got everything else right but went on to tie the local summer heat wave to the greenhouse effect in a more unique way than indicated by Hansen's graphs of probabilities and his colored dice, which were specifically designed to avoid such misinterpretations.[3]

CHAPTER
10
The View from the Greenhouse

I think it's kind of a personal thing ... Do you want to be a scientist-activist? [That is] obviously a decision that Sherry [Rowland] made, and it's a decision that Harold [Johnston] made at one time ... whereas Mario [Molina] has been very careful not to become an activist. One has to decide what kind of scientist you want to be.   Dr. Susan Solomon, 1990

NCAR (National Center for Atmospheric Research), Boulder, Colorado, Friday, June 24, 1988

Dr. Stephen Schneider laid the newspaper on top of his crowded desk, shaking his head almost imperceptibly. Jim was in for it now, he thought, and the poor guy probably didn't realize yet how much of a fire storm his 99 percent confidence statement was sure to create, not to mention what the reaction would be in government circles to his remark about "waffling."

Schneider thought about phoning Hansen, but the lengthy list of other tasks ahead of him this Friday changed his mind. Just twenty-four hours ago he had been with the famed Dr. Edward Teller during a conference in high-altitude Flagstaff, Arizona, and in two days—on Sunday—he was scheduled to fly to Toronto for a World Meteorological Organization conference on the changing atmosphere. In fact, he thought, rubbing his eyes, he could feel the jet lag again. With such a full calendar it was a common malady for him these days, and thanks to the heat and the drought—and Jim Hansen—it would inevitably get more crowded. The media would be looking for him, too.

Global warming had already been the subject of a barrage of newspaper and magazine articles as the summer got hotter and the rivers ran lower. Certainly the subject wasn't new—Schneider himself had been working on it and talking energetically about it for fifteen years. But with another respected scientist like Hansen welding a connection between the greenhouse effect and *actual* global warming as a matter of cause and effect, the media would be in a feeding frenzy, and his secretary had already started to field phone calls. Unlike Hansen, his friend and former colleague at GISS, Steve Schneider was one of the few atmospheric scientists who regularly talked with the media. With his curly hair, youthful demeanor, and infectious enthusiasm, Schneider came across very well on television, radio, and in print interviews. Jim Hansen, however, was a novice at the game, and Steve wondered how Hansen would handle the onslaught of interview requests that had to be cascading into the GISS offices. Hansen was a soft-spoken, studious sort, and instant notoriety would come as a profound shock to him. Steve's memory of countless conversations with Hansen at GISS played across his mind for a moment. In 1970, as an excited twenty-five year old, Schneider had joined GISS as a visitor and postdoctoral fellow working on the emerging field of human impact on climate. It was a rather illogical jump, since his newly minted Ph.D. from Columbia was in plasma physics, but he had always wanted to do socially useful work, and finding a postdoc position in environmental science a few blocks from Columbia in New York City was perfect. During the day he would work with men such as Dr. Ichtiaque Rasool (who served as his mentor), and at lunch he could sit with experts such as Jim Hansen, telling Hansen the latest twists and turns of *his* evolving discipline while Jim would describe the intricacies of the radiative-transfer calculations he was doing on the planetary atmospheres of Venus and Mars. Steve had great respect for Jim Hansen, but they had chosen different paths. While GISS had provided a rather insulated scientific cocoon for Jim from which he had erupted into public consciousness and political controversy only through reluctance—and now, perhaps, naïveté—Steve Schneider had plunged headlong into the challenge of being a communicator-scientist, a self-chosen role that early on branded him as an activist, and even an "alarmist." It was also a role that seemed to raise the hackles of fellow scientists as surely as dogs growl at cats.

The image of the weather map room door at NCAR came to mind. When Stephen Schneider had moved to NCAR from GISS in September 1972, he had barely learned the location of the map room before leaving for a science convention in Baltimore, but it was the perfect place for a disapproving editorial comment—as he soon discovered.

Schneider had been thrilled to get an invitation to speak about man's modification of weather to the December meeting of the prestigious American Academy for the Advancement of Science. He was just a young postdoc in 1972—a "kid" in his description—allowed to address not just peers but masters of the game. His thirty-minute speech had gone well, and he had even thrown in a little humor at the end, a twisted version of a quote attributed to Mark Twain, "Nowadays, everybody is doing something about the weather, but nobody is talking about it."

The New York Times' Walter Sullivan was there and wrote an article about the subject of his talk, quoting the young postdoc and even using his comedy line. But the piece beat him back to Boulder, and someone had added a pithy comment to the clipping and posted it with undisguised disgust on the map room door. The large letters stamped over the clipping were unsigned, but they left no doubt that at least *someone* at NCAR was singularly underwhelmed by Dr. Schneider:

"BULLSHIT!" it said.[1]

As a new recruit in such an established and respected institution, Stephen Schneider was an interloper expected (apparently) to enter quietly and respectfully, hat in hand, mindful of the *experienced* scientists arrayed in all directions who spent their hours committing acts of science, rather than acts of self-promotion such as talking to reporters. In effect, at that point in his development as a scientist, the unspoken code of the NCAR hills dictated that he should speak when spoken to—and he had violated the code.

Not just violated it, in fact, but decimated it, getting himself quoted in *The New York Times* under Walter Sullivan's respected byline, and getting caught making a funny, insouciant, public remark to boot. It had all obviously exceeded someone's tolerance level within the quiet research warrens. The unsolicited peer review was a barometer of professional pique from the ranks.

Sullivan's story then began making the wire service/syndicated rounds of less careful newspapers around the country—losing details and coherence as it underwent the metamorphosis of rewrites and cuts. As it migrated, more clippings appeared on the map room door (though the midnight stamper did not strike again), each of them posted without comment, but posted all the same, which was the important point. Virtually no one at NCAR missed the fact that young Dr. Schneider was off to a rocky start.

A scientist lives and dies by his reputation, but for Stephen Schneider that caveat became a unique mixture of agony and ecstasy over the next few years as fellow scientists acknowledged the precision and worth of his research, yet decried his tendency to be vocal and verbal and visible to the press and lay public alike. His was a concept in collision with the classic view of how science should be done (quietly, and apart from the rabble), and his willingness to be open and communicative rapidly alienated many colleagues.

In adjacent fields, Schneider knew, there were similar parallels. Carl Sagan, for instance, now a professor at Cornell, was denied tenure at Harvard after being enormously visible in books, lectures, and television. The ire of scientists who have toiled for decades without limelight and international notoriety can be visceral and vicious, especially when fellow scientists seem to be setting the standards by media exposure rather than contributing to the peer-reviewed scientific process. In the late seventies, for instance, bulletin boards in a plethora of science departments suddenly bore clippings of a particular newspaper article that seemed to crystallize the problem—it was a pedestrian piece on space travel by a reporter who had at some point interviewed Carl Sagan, and was obviously impressed by the experience: "According to Dr. Carl Sagan," the reporter wrote, "the speed of light is 2.99×10^8 meters per second," as if one of the most basic numbers in science was somehow in dispute.

By the late seventies, even the doctoral students who came to NCAR to study under Stephen Schneider were feeling the heat. Because they were students of a renegade, they were subjected to extensive questioning and other subtle forms of harassment when they returned to their home universities for their final oral exams (or when such postdocs sought positions with other research facilities). That sent a crystal clear message: "Those of us who hold the torch of *real* scientists are suspicious of anyone exposed to the heretical concepts and ideas of this overly communicative maverick Schneider."

In other words, by implication, "real" scientists should speak softly and only to each other, especially where policy matters are concerned.

"It's your fault, Stephen," Schneider was told by no less a personage than Dr. Margaret Mead—whom he had come to know in 1975. "You were warned what the reaction would be before you started being so visible." You have to learn to transcend it, she counseled, develop a thick skin, and make sure your science is impeccable so that they can't criticize you on that, then clearly separate out the facts and the values. Margaret Mead knew from long experience that he was attempting to engage in interdisciplinary science by discussing science policy in probabilistic terms, and that, quite simply, was deemed a technical violation of the scientific culture.[2]

But in June 1988, the irony of the moment was rich, and it wasn't lost on Stephen Schneider for a second. Here was the balanced, nonradical director of GISS, Dr. Jim Hansen, saying his confidence level was 99 percent that the world temperature averages for the last three decades showed that the greenhouse effect and global warming had been detected.

Yet Steve Schneider, who to some appeared to be a radical alarmist, would not have gone that far. To Schneider, an 80 percent confidence level would be the maximum appropriate figure, and even then, if he had been in Hansen's hot seat before a Senate committee he probably would never have made the flat assertion that the greenhouse/global warming had, in fact, been detected. That was an antagonistic statement to a large number of colleagues who saw no indication that the rising temperature curve had yet lifted out of the noise level of natural climate variability.

Steve Schneider, in other words, would have played it far more conservatively than James Hansen.

By the time Schneider reached Toronto on Sunday, the questions had started, many of them aimed at getting him to side with, or oppose, Jim Hansen. But Steve Schneider didn't *disagree* with the major conclusions Hansen made, merely with the extent to which his confidence level could be expressed numerically, and yet that sounded like bet hedging to the media. After all, they were finding a loud and upset reaction from other scientists (such as Dr. Richard Lindzen of MIT) rather easy to come by. For a well-known atmospheric scientist such as Steve Schneider to waffle on the point himself would be unthinkable. Surely he'd attack or defend. All the reporters had to do was find him—which they did in Toronto on Monday.

"Then you disagree with Jim Hansen?" asked a newsweekly magazine reporter for the third time.

"I'm not going to get into a false dichotomy debate with Jim Hansen," Schneider replied again, "... when we both agree that the physical basis of the greenhouse effect is very strong, that greenhouse gases have increased, and that global warming is occurring on the planet."[3]

One of the problems, Schneider knew only too well, was the fact that too many reporters simply did not understand the issue, and there were times he became quite weary of explaining it yet again. Having worked on inadvertent climate modification issues since the early seventies—long before "global warming" or "greenhouse effect" became household phrases—the deadly serious questions of excess carbon dioxide and trace gases in the atmosphere of planet Earth were second nature to him. But the American public was just waking up to the idea that the sky was not an inexhaustible, eternally forgiving public sewer, and the possibility that there might be thermal and physical atmospheric consequences to the "progress" of modern civilization was sizzling home with the vengeance of the summer's heat wave. Whether due to the heat, Hansen, or spontaneous consensus, masses of people were asking the right questions (through the media) for the first time, and as he had always felt a responsibility to do, Steve Schneider would try to provide some of the answers.

And as had been the case for so long, too many fellow scientists would watch or listen to Schneider's comments in the following weeks and feel a primal urge to deride or attack him with the same zeal reserved for a Sherry Rowland or a Carl Sagan—not for what he said so much as for having the temerity to say it: the unforgivable act of science speaking to the ignorant public through the garbled medium of the press. At least Jim Hansen appeared very uncomfortable in the television spotlights. But Steve Schneider actually seemed to *enjoy* the process, and that was heresy.

The words seem so simple: the Greenhouse Effect. Yet if you've never been in a true greenhouse, that can be a rather hollow phrase.

Entering a glassed-in greenhouse on a cold and sunny day is a visceral and somewhat heady experience, especially if the owner is a gardener of merit. The rich colors of blooming flowers, the fertilized, moist soil in the various planters, and the heavy, earthy scents carried on the wings of invisible water vapor permeate the enclosed airspace under an expanse of clear glass panels that form the walls and ceiling and create an indelible impression.

From the outside, a greenhouse looks lush and exotic. Inside, it *feels* exotic. One moment the sun is merely a hint of incandescent warmth on your hair and face as you approach through the chill air, a coat pulled close around you (with, perhaps, the profusion of autumn leaves or the starkness of snow-covered landscape providing a seasonal accompaniment to the crisp air), and within the time it takes to open the doorway and pass through, you're in a fragrant jungle. In late afternoon your coat comes off fast, for the earth tones and smells surrounding you hold both a heavy humidity and a smothering warmth that seems to radiate from the very molecules of the invisible gases that enfold you.

Indeed, that's precisely what's happening.

On a moonless night following a sunny day, were you to take a military-style nightscope (which can "see" infrared light, the electromagnetic energy found just below the wavelengths of visible light) and use it to look at the greenhouse from across a field, the structure would show up as a profusion of brightness, the radiant waves of infrared undulating from within. Only a fraction of the infrared inside such a greenhouse escapes, however, because the glass walls are opaque to infrared. Most of the black light—the heat energy glowing from the heated objects and molecules inside—tends to remain within, leaking slowly through thermal transfer to the cold night beyond, much as a thermos bottle slows the release of heat from a hot liquid inside.

The thermal energy inside the greenhouse, which creates the black glow to begin with, comes from the wooden boxes and the soil and the plants themselves that have heated up through the day by absorbing sunlight, but it also radiates from the molecules floating in the air trapped within the greenhouse, and therein hangs a tale.

When you stand inside a greenhouse luxuriating in the trapped warmth, it's obvious that the sun's energy has provided the heat, though not by heating the glass directly. The visible sunlight passes largely unimpeded through the glass, leaving very little heat energy in the molecules of the glass itself. But the sunlight causes the rich collection of molecules of the plants and soil, the boxes and gardening equipment, and everything else within to resonate with increasing heat energy—energy that gives off more and more black body radiation (infrared) as the temperatures of those surfaces rise.[4] As the stored heat increases, it emits more infrared light, much of which bounces off the infrared-opaque glass overhead. This infrared energy is kept from escaping, and in turn, it keeps the interior temperature warm.

Some of the gases inside the normal greenhouse are even more opaque to infrared rays than the glass walls and ceiling of the structure. Nitrogen (78 percent of the atmosphere) is not one of them, nor is oxygen (21 percent). Nitrogen and oxygen molecules (N_2 and O_2) are transparent to infrared and simply allow it to pass through. But gases such as water vapor (H_2O), carbon dioxide (CO_2), and methane (CH_4), resonate—vibrate and emit infrared light themselves—causing more infrared rays (carrying heat energy) to bounce back and forth, raising the temperature of everything within.

Most greenhouses, then, do a highly efficient job of trapping the sun's visible light energy—and the resulting heat—

inside, storing it in the plants and equipment as well as the greenhouse-sensitive molecules of the air within the structure itself. In fact, if there are no windows to open and close, the heat can become oppressive, eventually damaging the plants within.[5]

Yet the gases that provide that insulating service constitute only 1 percent of the atmosphere (not counting water vapor, which for the Earth at large can add anywhere from near-zero percent upward of 4 percent to the total volume of atmospheric gas that responds to infrared rays).

Earth's atmosphere, like the glass of a greenhouse, is transparent to the incoming visible light of the sun. But the Earth's surface and all the things upon it resonate infrared radiation, much of which is absorbed by the greenhouse gases in the troposphere and returned to the surface in a cycle that works like a porous blanket—porous because there are "safety valves" to keep us from overheating. One of those valves is simply the amount of greenhouse gas in the atmosphere. The greater the percentage, the higher the temperature.

If all the infrared radiation emanating from Earth were trapped—if greenhouse gases made up too great a percentage of Earth's atmosphere—the result would resemble Venus, where the majority of the atmosphere is a crushing blanket of carbon dioxide which has created a runaway greenhouse effect (temperatures at the surface are in excess of 700 degrees F, hot enough to melt lead). On Mars, by contrast, there is too little carbon dioxide in the thin atmosphere to trap infrared heat, and the surface, as a result, is colder than Antarctica, and (as far as we know) stone-cold dead.

This is the "Goldilocks Effect" (Venus is too hot, Mars is too cold, and Earth is just right). For decades it was widely believed the difference in the three planetary temperatures could be explained by their relative distances from the sun. We now know that's wrong—the difference in solar distance for each planet cannot account for the nearly thousand-degree (F) difference in surface temperatures from the closest to the most distant. If that *were* the only governing factor, Earth's surface temperature would be just a bit above 0 degrees F, and the surface of Venus would be less than 100 degrees F. Something else had to be found to explain the phenomenon, and the answer (which came with the first surface temperature data from Venus radioed back by spacecraft) is the presence, absence, or correct concentration of carbon dioxide (and trace greenhouse gases) in each atmosphere.

If the amount of greenhouse gas in an atmosphere determines how effective the heat-trapping mechanism is, then it becomes very obvious that linear increases in the concentration of greenhouse gases will cause less infrared radiation from the sun-heated surface of the Earth to escape back into space, and thus the entire ecosphere will retain more heat, eventually raising the average temperature that must be absorbed by the land masses and oceans of the planet. There is no question now of whether this relationship is valid and true. The debate is merely over *how much* additional carbon dioxide the Earth's atmosphere can absorb before the effects of the added atmospheric heating (because of the increasing opaqueness of the atmosphere to infrared) become apparent at the surface, and how much surface and atmospheric heating it will take to begin measurable changes in such things as weather patterns (climate), sea levels, and amounts of glacial ice. These latter questions have no certain answers as yet.

But go back for a moment to the word "linear." Carbon dioxide molecules trap infrared light, but not all of it. Each molecule has an "absorption spectra"—a range of electromagnetic frequencies that will *not* simply pass through the molecule, but will leave heat energy behind, or cause some other change. (CFC molecules, for instance, are broken apart by UV-B frequency radiation at very high altitudes.) By the same token, each molecule, including carbon dioxide, has "windows" in its spectra—frequency ranges that are not absorbed. In the case of CO_2 there are windows in the wavelengths of infrared light that allow some of the black body radiation of heat from Earth's surface to pass through and on out to space. Adding anything to the atmosphere that absorbs infrared in the same wavelengths as those so-called windows would obviously plug up one of the safety valves in the greenhouse, and increase the retention of heat between the atmosphere and the surface.

There are such molecular plugs, and their quantities are being increased by the actions of human beings.

Methane (CH_4), for instance (which is increasing inexplicably at the rate of 50 million tons per year on top of the existing annual emission rate of 500 million tons per year), is 20 times more effective in trapping infrared than CO_2. One molecule of methane absorbs as much infrared radiation as twenty molecules of CO_2. Nitrous oxide (N_2O), which is being added at the rate of 50 million tons per year, is 250 times more effective than CO_2.

But there is yet another class of molecules still being spewed into the atmosphere at record rates that—in addition to being a carrier of chlorine and a destroyer of ozone—is almost exponentially effective in trapping infrared radiation: CFC's. One single CFC molecule in the troposphere traps as much infrared heat energy as *twenty thousand molecules of CO_2*!

The Earth's atmosphere is massive when you deal with the numbers necessary to precisely describe its scope and breadth; whether in gigatons, cubic miles, or whatever other measure, the volume is enormous. But when you realize that life on this planet has been made possible by the infrared opaque qualities of permanent gases that constitute less than 1 percent of thin atmospheric envelope (plus the 0 to 4 percent range of water vapor in the air), and yet those relatively tiny amounts of gas raise and maintain Earth's temperature with such efficiency, the fact that humans have been belching massive quantities of additional infrared-resonant gases into the sky for the past hundred years becomes alarming.

Especially if you discover—as a Caltech scientist named

C. David Keeling did in 1958—that the gaseous mixture overhead really *is* being significantly altered.

The International Geophysical Year of 1957 (the IGY) was a great leap forward in science. The theory of Plate Tectonics in geophysics, for instance (which revealed that continents move continuously on crustal "plates" that produce earthquakes, volcanoes, and mountain ranges wherever they meet), was partially a dividend of research nurtured and perpetuated by IGY. More important, IGY seduced some of the best and brightest minds in science into their respective fields by providing tools, time, and funding, pushing our knowledge of the planet from shadowy postulations to sound theories supported by hard evidence gleaned from field research. The ranks of such IGY scientists include Joe Farman, who would finally validate in 1984 what NASA hadn't noticed: the Antarctic ozone hole; Frank Press, later the science adviser to President Carter and head of the National Academy of Sciences; and in one isolated but very important corner of chemistry, a young postdoc named Keeling, who had "an overwhelming desire to measure carbon dioxide."[6]

The basic elements of the greenhouse effect, and even the concern that atmospheric levels of carbon dioxide *might* be rising with unknown consequences, were not new concepts. In 1827, for instance, a French mathematician named Jean Baptiste Fourier (primarily remembered for his work on the theory of thermal transfer) realized that the Earth's atmosphere retained the sun's heat to keep the surface habitable—characteristics he described by comparing the atmosphere to a common greenhouse. But it was nearly 1860 before a British physicist, John Tyndall, took the process of deduction a step further by looking into *how* that greenhouse operated. Tyndall discovered that *some* atmospheric gases specifically absorb heat radiating back up from a sun-warmed planetary surface: carbon dioxide, water vapor, and ozone, to be precise (although there are other natural trace greenhouse gases, such as methane).

The Industrial Revolution was in full swing at the end of the nineteenth century, before anyone formally linked man's release of carbon dioxide to potential climate change—and even then, the implications of global warming weren't all apocalyptic. Swedish chemist Svante Arrhenius was the first to formally realize that massive burning of coal and wood was releasing literally thousands of tons of carbon dioxide into the air each year, and if carbon dioxide was a greenhouse gas and the mainstay in Earth's thermal blanket against the coldness of space, then increasing the thickness of that blanket would logically tend to raise the temperature of the surface, perhaps to dangerous heights. Then again, there might be a positive side, wrote Arrhenius in 1906. A warmer atmosphere might mean better crop production all over the world. And, he said, if the amount of atmospheric carbon dioxide was the governing mechanism for the Earth's ice ages and glacier advances, a rising CO_2 might prevent a return of the glaciers in the distant future.[7]

Arcane scientific musings, however fascinating, seldom galvanized public attention in the age before mass global communication. Even within the scientific community, over a half century would pass before the possibility that CO_2 might be *accumulating* in the atmosphere gained wide acceptance. After Arrhenius's observations, those geoscientists who bothered to pay attention assumed the products of combustion—including CO_2—fell back out of the atmosphere, either into the vast oceans of Earth, or back to the soil in some form. After all, within every seashell of calcium carbonate, and every tree trunk full of trapped carbon (trees breathe in CO_2 like all green plants, and give off oxygen as a by-product) there was ample evidence that Earth had the systemic ability to balance the equation. And since things weren't getting any hotter at the surface, any real concern over global warming as a result of the greenhouse effect was nil to none.

That lack of concern remained until 1957, when oceanographers Roger Revelle and Hans Suess discovered that the Earth's oceans resist taking in extra carbon dioxide.[8] Suddenly, with that information, the CO_2 equation seemed off-balance.

The "Revelle effect" discovery damaged the confidence of earth scientists that the planet could shrug off any changes in CO_2 concentrations wrought by man's activities. As with the CFC controversy decades later, the question began to creep into the back of several minds: Where does the additional CO_2 *go*, if not in the oceans?

Then came David Keeling, a 1954 chemistry Ph.D. graduate of Northwestern who wanted to go west where he could pursue a scientific career and backpack at the same time. Joining the geology department at Caltech, he rapidly found that measuring carbon dioxide (which required taking outdoor samples) might be a useful project. It would turn out to be his life's work.

As his first years at Caltech passed, Dr. Keeling wasn't terribly interested in the consequences of *finding* an abundance of CO_2 in the tropopause; he merely wanted to *measure* it. While producing (at a snail's pace, according to his mentors) new and refined instruments to make incredibly precise measurements of the gas in every conceivable place, Keeling's first research project rapidly metastasized into a worldwide, thirty-year obsession.

In 1957 Keeling jumped at the chance to join the IGY effort with a grandiose project to measure CO_2 levels all over the world—including such lofty places as the summit of Mauna Loa volcano on the big island of Hawaii in the state of Hawaii. Mauna Loa stands 13,677 feet above the tropical Pacific Ocean, high enough to be snow-covered in the winter, and to provide a precious commodity: unsullied air. From the weather observatory near the summit, Keeling found his instrument could snatch an amazingly pure sample of blended, averaged atmosphere unadulterated by the local influences of plant and animal (and human) life.

David Keeling had already tracked the daily cycle of CO_2 concentrations in a forest, watching the levels rise during the night (when no sunlight was around to cause the photosynthesis process that uses CO_2 and gives back O_2), then fall to a startlingly consistent 315 parts per million every afternoon. Those before him who had tried to measure CO_2 had indicated that the concentrations varied wildly, averaging 300 parts per million, but dropping and rising with the locale. Keeling, however, found a global average—a basement figure. When he began to take samples of air at the top of mountains such as Mauna Loa, he also found *those* readings bottomed at 315 (although he discovered that the worldwide average changed with the seasons to a high, at that time, of 318 in late fall to a low of 315 in the spring—the lowest global concentrations of CO_2 occurring in spring because the level of photosynthesis in the northern hemisphere means plants are growing and using CO_2 at the highest rates).

Keeling wasn't the first to measure carbon dioxide, but he was the first to realize that there was a global atmospheric average—and that the average was changing. His spring readings taken at the top of Mauna Loa were rising year by year. When graphed (the concentrations of CO_2 in parts per million versus time), it was like tracking the vertical position of a small boat bobbing gently on the swells of a rising tide over a period of hours. The motion was up and down annually, but the overall trend was ever higher.

In other words, at least a substantial portion of the CO_2 being added each year to Earth's atmosphere was *not* going into the oceans or back on land, it was staying in the atmosphere—measurably.

While Keeling's first readings had a low of 315 parts per million, thirty years later in 1987 they would be approaching 350 parts per million, and more ominously, the rate of increase would begin to grow. In science, even a straight line is called a "curve" when it traces multiple data points in search of a trend, so this tendency of CO_2 to continue to build up in the atmosphere has become known almost universally in earth science as "Keeling's Curve."

But is the rise in atmospheric CO_2 concentrations truly an *unprecedented* change, whatever the consequences? That raises historical questions for which there were no historical answers back in 1958: What was the atmosphere like two hundred years ago? What was it like a thousand years ago? Was it possible that instead of an anthropogenic effect, the rise of CO_2 concentrations in the atmospheric envelope could be a long-term variation that would fall eventually as surely as it had risen? Such questions failed to take on any great momentum or international significance as the sixties progressed, but they remained there in the back of many fertile minds like a nagging feeling of something left undone. Without a historical record of CO_2 concentrations stretching back at least a few hundred years, there was no way to guess the answers.

Until, that is, one of the IGY expeditions uncovered an ancient library that had been frozen in the depths of Earth's ice caps for thousands of years.

It was an American team of ice experts that made the breakthrough. They were trying to learn more about permanent glaciers by drilling into the thick ice sheet that covers Greenland's northwestern flank, extracting the ice core from their drill hole for laboratory examination. There had been theories that in ice-capped regions of the Earth, core samples pulled from the ice might yield identifiable layers. If that was true, and if they *could* count backward through the layers season by season, ice cores could provide a stratified record of snowfall and ice depth that could be read much as the petrified silt at the bottom of lakes and oceans that have been compressed into sedimentary rock over the ages can be decoded as a sequential diary of Earth conditions. With the Greenland core samples, the theories turned to fact. There, layer by layer, were samples of snow flurries that had fallen a thousand years ago, blizzards that had occurred when Christ was born, and even traces of volcanic ash that had filtered down from upper atmospheric winds from the eruption of a volcano in the Aegean Sea around 1645 B.C.[9]

In addition, by carefully analyzing different features and factors—most of them invisible to the naked eye—even the concentrations and point of origin of pollens could be pinpointed, marking the seasons; and, by indirect techniques, the Earth's average temperature could be deduced.

But even though the realization came slowly, the scientists involved finally realized that within each square millimeter of ice core was also an invaluable gold mine of atmospheric information—a genuine library of preserved atmospheric data.

The ice core revealed something far more important than the thickness of snowfall and the relative intensity of winter—it revealed the very nature and concentration of ancient air. In each sample of ice were tiny bubbles of air that had been trapped between snowflakes and eventually compressed into ice-walled time capsules. There wasn't much air to work with in each bubble, but then when you're measuring parts per million or parts per billion, given the size of the gaseous molecules to begin with, it was merely a matter of building and refining the equipment to extract the tiny air samples and measure the molecular content. An entirely new branch of science had been discovered in the ice cores of Greenland—paleoclimatology, the science of deciphering past climatic conditions by studying the ice core records. Not by *inferring* what the air was like, but through analysis of the actual ancient air samples themselves.[10]

Greenland's ice cap is thick, but the ultimate library—the Library of Congress of ice core sampling—was as far to the south as you could get in Antarctica. There any atmospheric concentration will be truly representative of global averages since the location is so far removed from local influences. The Antarctic ice cap—especially the two-mile thickness of ice over the geographic South Pole—began forming perhaps five hundred thousand years ago. That record had to be tapped, though the methods and the money for doing so would take international cooperation and quite literally decades of work.

Through the sixties and seventies various competing research centers punched away at parts of the ice caps in Greenland and Antarctica, and even the ice fields of the Swiss Alps, learning how to build the right analyzing equipment and how to make the best use of the ice cores pulled from the depths. By the 1980s, core samples from the last few hundred years had been recovered and analyzed, thus finally beginning to fill in the picture of CO_2 concentrations before David Keeling conceived his curve.

In the early part of the eighteenth century the worldwide concentration of CO_2 was about 280 parts per million. By the time of the American Revolution, the concentration had begun to rise. Scottish inventor Sir James Watt had perfected his design for a steam engine, and the burning of coal was increasing sharply in Europe—and showing up at the South Pole.

The paleoclimatic breakthrough, though, came in the late seventies with the drilling of an extraordinary mile-long hole at the Soviet Antarctic outpost known as Vostok. As the drill descended, it cut through layers deposited more than 150 thousand years ago. Slowly and gently the five-thousand-foot ice core was pulled from the shaft, cut into sections, and flown (by U.S. aircrews, in fact) to a Swiss glacial research laboratory that had become known as the world's best at deciphering the ancient ice. Just as with the mountains of digitally encoded satellite data, the process of testing, reading, and deciphering the rich climatic record in those ice cores took several years, but when the results began to emerge as a coherent picture of various measurements through the ages, it was as if some of Earth's vital signs had been found on an old chart, running the length of 160 thousand years!

And the graphs of the average temperature curve and the average CO_2 concentrations in the ancient air samples provided a profound shock: They matched. There was a strong and undeniable correlation between the levels of CO_2 and the average global temperatures over the entire period of the past

160 thousand years: Whenever the curve tracing Earth's average temperature goes up, the curve tracing the average CO_2 concentration goes up almost identically. And when one falls, the other falls. Arrhenius's ideas of eighty years before, that CO_2 and the ice ages might be linked, suddenly seemed prophetic. In the last 160 thousand years there have been two major worldwide ice ages that resulted in mile-thick ice sheets that covered both northern Europe and North America as far south as New York, Chicago, and Seattle. Known as "glacial" (versus "interglacial," which means between ice ages) periods in Earth's climatic history, the last one ended about 10 thousand years ago, but was at its height 20 thousand years ago. The previous glacial period ended around 130 to 140 thousand years ago.

One hundred and forty thousand years back the Earth's average temperature stood nearly 4°–5° C lower than today's average, and worldwide CO_2 levels averaged about 195 parts per million.

As soon as that ice age ended, however, average temperatures in Antarctica shot up perhaps as much as 2.5° C higher than today, with the CO_2 concentration curve matching the rise almost exactly and topping out at just under 300 parts per million.

As the world's temperature curve begins to drop again during the interglacial period, the CO_2 curve for Antarctica drops right along with it, and in the midst of the last ice age, the average temperature of $-10°$ C (from today's average) once again corresponds very closely with the CO_2 average of just under 200 parts per million.

Considering where the data came from—actual samples of trapped air from each ancient period—the correlation between the two curves is amazing, and one of the significant discoveries of the twentieth century in earth science.

But just what do we have here, a chicken or an egg? Does the concentration of CO_2 in the atmosphere drive the average global temperature, or vice versa? There is no question now that where one reading goes, the other follows, but who's in the lead?

That remains one of the unanswered questions. We know *what* happened, but we don't fully understand *why*—and the answer to "why" could solve many of the present questions about what our massive alteration of CO_2 levels may mean for Earth's near and long-term future. The global atmospheric levels of CO_2 have risen 25 percent in the past century. Logic and the mechanics of the greenhouse effect indicate that the average temperature will follow, but does the historical record prove that, or merely add to the mystery?

Washington, D.C., Thursday, August 11, 1988

Dr. Steve Schneider was used to the routine. Drive the thirty-five miles from Boulder to Denver's Stapleton Airport, catch a direct flight to Washington National Airport, jump in a taxi at curbside and head for Capitol Hill. This time, though, the hearing he had been asked to attend held a bit more significance than the average. The same Senate committee that had elicited Jim Hansen's explosive statements on June 23 was now considering a bill to control Global Warming sponsored by Senator Tim Wirth.

The "National Energy Policy Act of 1988" would be the first law to directly address the problem, if it had any chance at all for passage *in* 1988, which it did not. With President Reagan still in the White House and the Hill full of doubting legislators constantly being buttonholed by worried industry lobbyists who wanted more study and no action, the bill was essentially dead on arrival—but it was a place to start.

And it gave Tim Wirth another opportunity to fuel the public consciousness of the global-warming issue through heavy media coverage.

The summer climate through most of the nation had remained brutal since June 23, and the various professional clipping services around the country were wearing out scissors trying to keep up with the snowstorm of articles on global warming, sea level rise, agricultural dislocation, and other atmospheric issues. Jim Hansen's "99 percent confidence" statement was liberally repeated in most such articles as a benchmark point of view against which all other opinions were to be measured.

Too many reports misunderstood the connection that Hansen made between the drought of 1988 and global warming. He said that the greenhouse effect increased the likelihood and severity of droughts, and he said that the greenhouse effect and global warming were already happening. Many people, however, interpreted that to mean that *the drought of 1988* should be blamed exclusively on the greenhouse effect. If that connection of faith were to be broken—if someone came forward at a later time to say that the summer of 1988 was *not* brought to you in thermal excess, courtesy of the greenhouse effect—the efficacy of the greenhouse effect itself would be questioned. Steve Schneider was well aware of the danger, as were the other two senior scientists who would testify with him in a three-man "panel": Dan Albritton, director of NOAA's Boulder Aeronomy Lab, and NASA's Bob Watson.

To Watson, in fact, Hansen's "99 percent confidence" statement imperiled the global-warming case the same way the CFC–ozone destruction connection might have been discredited if the Antarctic ozone hole had been blamed on chlorine chemistry—then found to be natural. If a cause other than global warming were found to be responsible for the drought of 1988, the same disaster was likely to befall the greenhouse, global-warming theories.

Watson basically agreed with most of Jim Hansen's statements, except for the belief that a cause-and-effect relationship had already been established. Now he and Albritton and Schneider had to help repair what they perceived as damage.

Steve Schneider took his seat in Room 366 of the Dirksen

Senate Office Building at 2:02 P.M. as Tim Wirth got the hearing under way with an introductory slap at the administration:

"Unhappily, it is difficult to find evidence of leadership from the current administration. Just last week DOE's assistant secretary for the environment, safety and health was quoted in the *Coal News* and in the *Energy Daily* as saying, 'I think it is folly to try to come up with a solution to a problem before we know what the problem is.' Unfortunately, the pattern of the administration's response to the global warming issue is well summed up in these quotes."

The associate undersecretary of the Department of Energy, Donna Fitzpatrick, was the first witness, and as expected, she emphasized the uncertainties and potential costs of acting before the science was certain:

> ... we believe that the action of the type called for in S. 2667 [Wirth's bill] would impose significant costs on some sectors of the economy, both within the United States and worldwide. Without further scientific understanding, we cannot be assured that these actions will be either effective or efficient.

To Steve Schneider, the role he had accepted and practiced so often was less that of an activist or advocate than that of a communicator, and this would be a communicative challenge. Only Senator Wirth sat before him in the hearing room (other than staff members)—the rest of the committee having been detained in the Senate chambers in a defense appropriations fight—and Schneider knew most of those senatorial minds were already made up one way or another. But for those on the committee, the staff members, and the media who *would* pay attention to what he had to say (in person or in the hearing record), it was important to translate the scientific realities in clear and unambiguous terms.

> I would like to state at the outset that the greenhouse effect as a scientific proposition is not controversial at all. In fact, it is probably the best accepted scientific theory in the atmospheric sciences.

Schneider spoke of the Goldilocks effect and the contrast of Earth's comfortable temperature with those of Mars and Venus, all the result of variations on the greenhouse theme.

> We in a sense owe our present biological existence to the greenhouse effect.
> So where is the controversy? The controversy is over whether an increment to the greenhouse effect on Earth from human activities is going to make the planet 1 or 5 or 10 degrees Fahrenheit warmer in the next 20 or 50 or 100 years.
> You can decompose this problem of the greenhouse effect into several stages. The first stage is how much greenhouse gases, carbon dioxide, methane, nitrous oxides, fluorocarbons, are people responsible for putting in the air. We know this is virtually a certain fact that something like 25 percent more carbon dioxide is in the air today than it was before the Industrial Revolution. And we know that because the ice masses in the poles, gases have been trapped from those dates and one can simply analyze the composition of gas bubbles trapped in ice cores.
> The second question is ... how does nature dispose of it? Right now ... of the carbon dioxide something like half of what we inject seems to disappear from the system.
> So where does that half go? The bulk of people think it goes into the oceans. Some think the trees

are going to grow a little faster because CO_2 enhances photosynthesis.

Schneider paused, and added that there was general agreement that if the current amount of CO_2 increase per year continued, the amount of atmospheric carbon dioxide would double in fifty to one hundred years. If that did happen, he said, computer climate models would be the main tool atmospheric science would have to use to predict the results.

> There is a whole variety of [climate models] run both in this country and in many other countries ... for over 20 years. They are all in rough agreement ... that if CO_2 were to double and we had a long enough time to wait for the system to come into a new balance, the global temperatures would rise on the order of 2 to 5 degrees Celsius.
> Now, CO_2 is not the only gas of interest. There are these other trace gases I mentioned ... and if ... you add them together, they [effectively double] the CO_2 amounts. They are about equally effective as CO_2.
> So, we could be seeing temperature changes in the next 50 to 100 years as much as 5 or more degrees Celsius if you want to believe what these models do.

Tim Wirth interrupted: There was a floor vote called in the Senate and they had to recess for a few minutes, leaving Schneider's argument suspended almost in midsentence.

When Wirth returned and asked him to continue, Schneider outlined with great care the possible effects of a global warming of just 3° C, emphasizing that the main effect was the changing of the probabilities—recoloring the faces of the climatic dice so they came up with warmer-than-normal summers more often than before. He tried to undo the damage he believed Jim Hansen's statement had inadvertently caused in leading the media to conclude the hot summer was being blamed on global warming, but that even in the best-case scenario, global warming could be massively disruptive to the world's population and economy, and Americans were surely not going to like the results—especially if the breadbasket of North America moved north to Minnesota and Alberta.

> Finally we reach the question of what we should do about all this.
> There are three categories of policy response: mitigation, adaptation, and prevention.

There had actually been proposals to use high-flying jet aircraft to fill the stratosphere with some sort of soot to block out sunlight and counteract warming, Schneider said. That would be mitigation, but the cure could prove more disastrous than the malady. No one had any idea what the long-term effects of such a scheme would be, and few responsible scientists were interested in finding out.

> The second category is adaptation. It is the one that tends to be favored most by, to use a word, economists. At a recent meeting an economist got up following a talk that I and a number of others had given and suggested that the greenhouse effect issues ... are of a very serious concern, but until we can give specific details of exactly where it [will be] wetter and drier and exactly how much extra air conditioning loads we will have and exactly what happens to agriculture, they can't apply cost-benefit analysis to figure out the winners and losers. So, therefore, the appropriate action was to study.
> And I recall saying that I was sorry. And I guess I wasn't being facetious; I was being nasty.

I was sorry that he wasn't a principal advisor to our military and the Soviets because, since we don't know from which part of the compass the missiles may fly or where the next terrorist incident is, we could spend $3 billion on studies and save $300 billion on hardware and on armies. Most people on the other hand do believe in strategic protection against plausible futures that we don't like.

The argument that I believe is more appropriate for a problem, such as the greenhouse effect or also ones such as the fluorocarbon stratospheric ozone that Dr. Watson will talk about [today], is to recognize the magnitude of the change that we are bringing relative to natural changes ... we are in a sense insulting the environment much faster than we can understand it. And therefore the policy issue is: are there things we can do to slow it down ...

The single most obvious one ... is energy efficiency. Energy efficiency does many things. It doesn't just simply reduce the greenhouse effect, slow it down, it also reduces acid rain. It reduces local air pollution crises in cities and populated areas, and it reduces dependence on foreign supplies. At the same time, it makes the energy component of our manufacturing costs lower and, indeed, the energy component of U.S. manufacturing costs is higher by perhaps a factor of two than many of our European competitors and even more than that relative to the Japanese. So in the long term it can make us more competitive.

But like any insurance, there is a premium, and that premium does have to be paid up front.

But this isn't simply a group of placard-carrying nutty scientists. There is a large consensus that there is a very rapid change relative to nature afoot. And there are mechanisms to slow it down, whereas even if this whole thing proved to be what I once called an infrared herring...

Schneider had to stop for a few seconds to let the laughter in the room die down.

...we would still have the benefit from these kinds of strategies.

In many respects that was the bottom line—the "classic" argument as Bob Watson described it: The immediate steps that nations like the United States can take to reduce energy consumption by increasing energy efficiency have great inherent value even if there is no greenhouse–global-warming problem. As Watson had explained it on several occasions:

"If we do these things, we won't look back in twenty years if there is no global warming and say, 'Oh, how awful we took these steps, because the greenhouse effect was a dud.' We'll be saying, 'How smart we were to do these things regardless of the greenhouse effect.'"

When it was Bob Watson's turn at the witness table, he walked out on his usual diplomatic tightrope to talk about what NASA was doing with a minimal research budget, while Tim Wirth tried repeatedly to get him to acknowledge that most of NASA's money had been diverted to the space program and so research into atmospheric problems was being grossly short-changed. Bob Watson, a NASA scientist in the delicate position of creating consensus and communication across disciplines and interests, could not make such a statement. Finally, cornered, Watson gave a seemingly vague reply, which, when read very carefully, turned out to be an eloquent and excellent example of the struggle to state the science without recommending the policy:

> Whether or not one should have a larger program, I think one could quite clearly state that there are a significant number of scientific uncertainties, and more money in earth sciences would not be misspent.

Bob Watson could not and did not verbally validate Senator Wirth's premise that NASA and the administration were scandalously shortchanging the funding of atmospheric studies, but neither did he rise in opposition to Wirth's essentially correct conclusion. Instead, he left the policy decision to the policymakers.[11]

One of the shared worries among Hansen, Schneider, and many others involved in climate studies was the inadequacy of the computer models to take into account the effect of the oceans in providing more water vapor which could cause more clouds to reflect sunlight (thus causing more cooling). Such feedback mechanisms might counter the effects of atmospheric warming. In even the most sophisticated supercomputer climate programs, clouds are treated as huge, square masses instead of individually shaped puffs. In the real world, however, clouds introduce an almost infinitely changing factor to the equations.

"Feedback mechanisms," as the ocean evaporation and changing clouds are called, are a poorly understood factor in the real climate. Yet, as the atmosphere heats up, the gigantic amount of water in the planet's oceans will not immediately follow a rise in air temperature. Does that mean the oceans will retard the global rise? Or by absorbing additional heat within vertical water circulation, will the additional atmospheric heat simply disappear in the depths of the oceanic aquasphere as in an unbelievably large heat sink? On the other hand, if the oceans do warm, couldn't that warming massively increase the evaporation rates (as well as raise sea levels since water expands when heated), thus increasing the concentration of water vapor in the troposphere, which—since water vapor is a greenhouse gas—could reinforce and intensify the global warming? Or would the increased evaporation rate mean more clouds deflecting more sunlight, thus *slowing down* the global warming? No one knows yet, and that's an important point.

If there are chinks in the armor of the computer models as a whole, this is where they're found: feedback mechanisms. Naturally, to those who for whatever reason do not want to believe that action is required immediately, this uncertainty becomes fertile ground for seeding more study in lieu of action. In fact, Steve Schneider's and Jim Hansen's most vitriolic critic is a scientist who claims that Earth's feedback mechanisms will ameliorate (if not eliminate) the greenhouse effect. Dr. Richard Lindzen of MIT, who has not used advanced computer modeling to arrive at his minority conclusions, has nonetheless loudly crossed swords with the majority of his colleagues by claiming that the Earth's oceans would moderate any global-warming effects to mere fractions of a degree Celsius. (Lindzen of course acknowledges the greenhouse effect, the greenhouse gas buildup, and the inevitability of some global warming, but debates how much warming will occur, and the severity of its effects.) Lindzen has few supporters who would downgrade the greenhouse issue to the degree he does, but as a member of the National Academy of Sciences and a respected scientist, he has influenced—and been quoted by—many of the reporters who perpetually hunt for a "balancing point of view," a process that in itself is often misleading.

Good scientists are required to state not only the limitations to their theories, but their opponents' arguments as well, and Bob Watson was not about to shirk that duty, especially in a congressional hearing:

> We [need] to understand . . . feedback between water vapor and the climate system. The amount of

ice will change globally, especially sea ice. A critical one is understanding clouds. It might be surprising, but even a small change in clouds or the altitude at which clouds appear could have a significant effect on the sign of the feedbacks between clouds and the atmosphere.

We need to understand basically the thermal response to the ocean. It isn't only the magnitude of the climate change that is important, but it is the rate at which climate changes. Until we understand atmosphere-ocean interaction, we are clearly not going to be able to quantify as well as we should the rate of change of atmospheric temperatures.

Finally, there was the underlying concern that had haunted all of them since the ozone hole had blown away the complacent idea that the atmosphere was a purely linear animal.

... Could the climate flip from one state to another? I think there is a very important message in Antarctic ozone that we should not ignore. We had a very aggressive program, we being the U.S. Government, to try and understand atmospheric ozone for more than a decade now. We thought we understood to a fair approximation what controlled atmospheric ozone. Unfortunately, we were caught by surprise. The Antarctic ozone phenomenon was completely unexpected and completely unpredicted. Antarctic ozone changed quickly and unexpectedly. I think we have to be concerned that the climate system may not change linearly, may not change slowly. It could jump from one metastable state to another.

It was obvious to Steve Schneider that the fight had just begun, and Senator Wirth's bill had no realistic chance of rapid passage—especially without heavy White House support. But he knew they had to force the political machinery to *make* a policy decision, not adopt one by default. As he had written in his formal statement:

> It is a cliché that "not to decide is to decide." The National Energy Policy Act of 1988 tries to confront directly the multifaceted elements of atmospheric change; it recognizes that unprecedented changes are plausible and that high-leverage actions can be implemented now; it does not use platitudes about scientific uncertainty to evade the need to act now. This proposed legislation is a good initial platform from which to build the national and international responses needed to help us cope with the rapidly changing atmosphere. It is often said that we do not so much inherit the world from our forebears, but rather we borrow it from our children. It is increasingly urgent that we act to prevent our debt from growing out of control.

Unfortunately, the next roll of the dice—and another pivotal research paper—were about to conspire to undermine all their efforts to clearly communicate the problem. Jim Hansen's words were about to backfire at a higher order of magnitude.

CHAPTER
11
A Time to Cry Wolf

New York City, July 14, 1989

Jim Hansen leaned back in his office chair and away from the computer keyboard, contemplating the words he had just punched onto the screen. He was upset, but the words and phrases of the letter he was writing back to *The New York Times* were reasoned and professional—and far more responsible, in his view, than the editorial *they* had run, which had misrepresented, and mischaracterized not only his statements and beliefs, but the entire greenhouse–global-warming issue.

It was a regrettable piece in Jim's estimation. It had shown up in the July 3 edition, prompted by the statement of an environmentalist who was never mentioned in the editorial—though Jim Hansen was. And it came in the middle of a relatively cool summer in the United States after a tumultuous twelve months that had seen unremitting attacks on the greenhouse effect, global warming, and anyone who dared warn of future climate change.

Under the banner of "The Editorial Notebook" and the title "Crying Wolf in the Greenhouse," editorial writer Nicholas Wade had alleged that environmentalists in general were wishing for another "horrible summer" in order to prove their points. Wade had not intended to paint Hansen with the same brush or allege that *he* wanted a scorching summer, but that was indeed the impression a reader could get.

In Jim Hansen's case it was untrue and unfair. As he had said many times, "I am not an environmentalist, I am a scientist."

Wade had written:

> Who could want to endure another summer like last, with drought-stricken crops, fish kills and ozone alerts? Some environmentalists, that's who. They hope it will "galvanize political opinion" and make Congress pass legislation to combat the greenhouse warming of the Earth's atmosphere.[1]

Had it really been an entire year, Jim wondered? He still found it difficult to recapture his research routine after the tumultuous publicity over his remarks at that seminal June 23 Senate hearing—the public hullabaloo that had transformed his GISS office from a reasonably quiet place to a madhouse of professional harpoons and the nonstop media requests that had frazzled his secretary, Carolyn Paurowski. Never mind Wade's wolf metaphor, it had been a bear of summer that held the nation in its teeth in July 1988, whatever the cause.

This year, however, July 14 marked approximately the halfway point through a relatively normal summer, but the absence of another heat wave was having a predictable effect: Those who had misunderstood his statements in the first place in 1988 were now concluding that since the summer heat was at normal levels and there was no drought, that meant the scientists were wrong, Jim Hansen was wrong, and the greenhouse effect was a fraud. Worse, scientists *like* Jim Hansen were nothing but . . . but . . . (sputter, spit, snarl) *environmentalists*! And *environmentalists* were all "alarmists" who as a class of individuals were all hoping the nation would be caught up in a second heat wave in 1989 because of the political effect it might have. Since there *was* no heat wave of 1989, though, it was apparently time to unmask the environmentalist attitude as irresponsible and counterproductive— which was precisely what Nicholas Wade seemed to be doing:

> Even another sweltering summer won't mean the greenhouse warming has begun. Droughts are part of the climate's normal variability. The danger in asserting they prove greenhouse warming has arrived is obvious.

Wade was right about that, but neither he, nor those outraged individuals who had rolled their eyes at the so-called doomsday warnings of ozone loss or global warming, seemed to want to acknowledge that Dr. Hansen had *never* tried to blame the summer of 1988 on the greenhouse effect. What he had tried to do was warn that the greenhouse effect would cause a global-warming tendency that would eventually load the climatic dice, making hotter summers and droughts *more likely*. And he had said that in his opinion the global signals had emerged from the noise—there was now a linkage of cause and effect between the 1° F of worldwide temperature increase and the greenhouse effect. The debate, of course, was over what climatic changes such warming might cause—an argument over *degree* and *timing* of change.

Nicholas Wade, who has a background in science, had been careful to state the science correctly. Yes, the greenhouse effect was beyond question. Yes, there was a substantial increase in CO_2 in Earth's atmosphere that many scientists *thought* might one day produce changes, *but* ...

> ... it is far from proved that the heat thus dumped in the atmosphere has yet caused a rise in global temperature.
>
> Many climatologists believe this is likely to happen, sooner or later. But they don't believe the half-degree [Celsius] of global warming in the last century can yet be attributed to the greenhouse effect. Nor do they see reliable evidence that the greenhouse warming has yet begun.

That was the crux of the issue. Jim Hansen felt the signals of global warming were beginning to emerge from the background of natural climate variability, but the majority of his colleagues seemed to disagree, at least in public.[2] Not even Dick Lindzen, though, doubted the greenhouse effect, or the theory that atmospheric warming to some degree has to occur when the amount of CO_2 in the atmosphere is increased. In fact, the vast majority of climatologists agreed with the basic concept that with enough added CO_2, eventually some global surface average air temperature rise *would* occur. What they mainly differed on was twofold: (1) The resulting global temperature rise would cause dramatic climate changes (sea level rises, desertification of agricultural areas, intensified storm damage, etc.); and (2) There was *already* an established, readable, provable cause-and-effect link between acknowledged global warming and worldwide average temperatures.

But neither Jim Hansen nor anyone else who could wear the title of responsible scientist had said that the heat of the summer of 1988 was the direct, smoking-gun product of the greenhouse effect. Yet, the implication radiated from the editorial:

> James Hansen of NASA is the only climatologist to have told Congress that the greenhouse effect is

> already here. Environment lobbyists are tempted to emphasize the threat of the greenhouse effect because it reinforces so much of their other agenda. If real, it adds a strong argument to their laudable push for energy conservation, fuel efficiency and preserving tropical forests.
>
> Crying wolf in the greenhouse will not work. Nor is it necessary. There is already an international agreement to prevent further thinning of the ozone layer in the high atmosphere, even though the predicted disaster—an intenser flux of ultraviolet rays from the sun—has not yet occurred. Over the last decade, ultraviolet radiation measured at the surface of the United States has, oddly enough, decreased. If the nations of the world can act to avert the ozone-layer threat before clear proof of harm, they will presumably do the same with greenhouse.[3]

To Hansen, the implications were intolerable, and Wade's faith in the ability of the international community to stop debating the merits of stepping off the tracks before the train came was naïve. It all added up to the same thing: Jim Hansen had to reply, and a letter to *The New York Times* editorial page was the proper format:

> A recent Editorial Notebook by Nicholas Wade ("Crying Wolf in the Greenhouse," July 3) leaves the possible impression that I have used a single hot summer as a basis for arguing that greenhouse warming is altering global climate now.
>
> During the drought of '88 I explicitly cautioned Congress not to blame a specific heat wave on the greenhouse effect.

He could have been whispering into the wind, for all the good that caution had done. By August it had been Jim Hansen against the world.

Hansen had lost a lot of sleep during July and August 1988 over the barrage of public and professional scrutiny. It was daunting to have one's own peers so strongly disapproving of a person's conclusions, but he had called it as he had seen it, and by fall, Jim Hansen was convinced that speaking frankly as he had done was the only scientifically honest course of action. He had seen a cause-and-effect connection between the past years of worldwide temperature increase and the global warming expected from the man-made increase in greenhouse gases. He had seen it, and he had talked frankly about it.

And he had no apologies.

In December the legacy of the public's misunderstanding came back to haunt him when a team of researchers at NCAR blamed the drought on a patch of cold water in the South Pacific. Kevin Trenberth and Grant Branstator had noted that a colder-than-normal sector of ocean off the west coast of South America could affect the weather in the northern hemisphere just as effectively as *El Niño*, the hotter-than-normal condition that can form in the same locale. They ran computer simulations with differing water temperatures in the southern hemisphere and saw events occur in their models that mimicked the summer of 1988: The jet stream over North America split in two, one segment bypassing the Midwest to flow over Canada, the other flowing over Mexico, leaving the central United States to bake.[4] Trenberth and Branstator specifically pointed out that their findings did not disprove any potential influence of the greenhouse–global-warming postulations, nor did they try to say that the drought was *exclusively* the province of the reverse *El Niño* event in the South Pacific. But those warnings were no more effective than Jim Hansen's repeated cautions than no one drought could be blamed on global warming.

The media pounced once again, missing the details and latching on to what seemed like another juicy scientific controversy. Hansen held one opinion, but here came two new scientists with an equal and opposite opinion![5]

"*The Greenhouse Climate of Fear*," screamed one guest editorial in *The Washington Post* on January 8, 1989, a piece widely reprinted around the country through the *Post*'s syndicated news service.

More hearings followed on Capitol Hill during February in direct response to the questions raised by such editorials and articles, which seemed to be dividing the issue of future effects from global warming into two simplistic camps: The greenhouse effect is a fraud because the summer drought of 1988 was caused by something else; or, regardless of whatever might have happened in the South Pacific during the summer of 1988, the greenhouse effect (and global warming leading to worldwide catastrophe) is here, and the sky, in effect, is falling.

Jim Hansen had weathered the entire period as stoically as he could, but editorials such as Nicholas Wade's simply went too far.

What was even more irritating in Wade's case was his reference to Jim Hansen's late spring battle with the administration censors.

Hansen leaned forward and began typing again.

> The article also misrepresents the nature of alterations by the Office of Management and Budget of testimony I delivered to Congress this year.

In April, OMB had finally gone one step too far. Scheduled to testify before the Senate Committee on Commerce, Science and Transportation (Subcommittee on Science, Technology, and Space), Hansen had routinely submitted his planned comments and waited in vain for OMB to return the marked up script.

Instead a phone call from NASA's headquarters instructed Hansen's secretary to send the testimony in via electronic mail to a word processor in Washington. Why, Hansen asked? So they can insert the OMB mandated changes themselves was the reply.

Soft-spoken Jim Hansen was furious. He called NASA Headquarters and literally yelled that OMB could *suggest* anything they liked, but any changes made to testimony under his name would be made in his office in Manhattan. As that message wound its way through the Washington bureaucracy, Hansen began circling the wagons and preparing to defend his script.

This was important testimony in his estimation, his first congressional statement since the infamous summer of 1988. It was a chance to get a more substantial discussion of the impact of the greenhouse effect on droughts in the record.

Hansen's group at the Goddard Institute had developed some very useful insights in the intervening months by studying the effects of warming on the global hydrologic cycle of evaporation, rainfall, and river runoff. The results were at once paradoxical and yet straightforward, and if they were correct, the implications for the planet were truly staggering. Their climate model showed that an increasing greenhouse effect led to a considerable intensification of *both* extremes of the water cycle: more droughts and forest fires on one hand, and more extreme rainfall, floods, and storms on the other. An increased greenhouse effect caused more heating of the Earth's surface and therefore more evaporation, especially over ocean regions, and this led to occasional episodes of more intense rainfall, especially moist convective storms, which raised the specter of stronger, more frequent hurricanes and more powerful thunderstorms dumping great quantities of water in short periods of time on land masses unable to accommodate such downpours without flooding.

But the model also showed that in times and places of relative dryness, the increased greenhouse heating and evaporation caused increased droughts and, by implication, increased forest fires. As the model progressed, weather patterns continued to fluctuate from place to place and from year to year as always, but the *chances* of having severe storms and floods, and the *chances* of having severe droughts and forest fires, both increased with an increasing greenhouse effect.[6]

A list of specific OMB–mandated changes finally arrived from NASA headquarters, and it was a mixed bag. First, Hansen was told to preface his testimony with the statement that it was only his opinion, and did not represent "a consensus of the scientific community." That was acceptable to Hansen, who knew well that such a consensus would take years to develop. Speaking in the voice of "my opinion" would give him the license to fully describe the GISS results, and their implications as he saw them.

The other changes, though less drastic than those OMB had imposed in 1987, were nevertheless hard to accept, especially a statement Hansen was supposed to add which was in the first person: "I must stress that [these results] should be viewed as estimates from evolving computer models and not as reliable predictions." The "not reliable" line was too much, but word came back from OMB that Dr. James Hansen would either swallow the changes in "his" testimony exactly as dictated by OMB, or he would not be allowed to testify as a government witness.

The alternative wasn't much better. Testifying as a private citizen would weaken the impact. Reluctantly, Hansen agreed, reserving the right to make his personal opinion clear verbally during the hearing. To make sure he had the chance, he sent a note to Senator Gore suggesting some leading questions the senator could use to ferret out which words were those of Hansen and GISS, and which were the handiwork of the OMB censor.

Gore was more than supportive, he was enraged at the political censorship. Al Gore and his staff began loading their guns for the hearing, and several days in advance revealed to the media and the world that the new "Environmental President" (as President Bush had styled himself in his campaign) was indirectly permitting not just muzzling of government scientists, but wholesale rewriting of scientific findings. Calling the rewrites a "form of scientific fraud," Gore excoriated Bush and his new administration while praising Hansen for his courage and warning with unabashed hyperbole that if the administration attempted "any kind of retribution in return for candor, they will have on their hands the congressional equivalent of World War III."

At the White House, a startled spokesman, Marlin Fitzwater, denied that Hansen had been muzzled by the changes that he admitted had been made, saying that Hansen's original testimony gave the impression that "there was unanimity within the government on [the global warming] matter, [and] we simply wanted to point out that there wasn't."[7]

Nicholas Wade had come to the administration's defense in his editorial:

> The Office of Management and Budget drew criticism when it heavy-handedly made Mr. Hansen qualify testimony to Congress last month. But its caveat, that his warnings "should be viewed as estimates from evolving computer models and not as reliable predictions," was scientifically impeccable. The computer models of the greenhouse effect are indeed "evolving"—they're somewhere around the amoeba stage.

Not surprisingly, Jim Hansen held a rather different view:

> Mr. Wade quotes out of context a single caveat which OMB inserted into my testimony: that climate simulations "should be viewed as estimates from evolving computer models and not as reliable predictions." If he refers to the climate of a specific region, the caveat is indeed "impeccable." But Mr. Wade, and the OMB censor, failed to note that my conclusions did not require precise climate simulations. Thus the function of the censor's caveat was to obfuscate my testimony.

Furthermore, this caveat was not the only alteration of my testimony. OMB also inserted a statement suggesting that current increases in greenhouse gases may be a natural phenomenon, and OMB altered my recommendation for steps which would reduce the greenhouse effect and make good sense anyhow: specifically [the] phase-out of chlorofluorocarbons, improved energy efficiency, and reduced deforestation. . . .

It is easy to denigrate "evolving" climate model studies as being "around the amoeba stage." But does Mr. Wade realize that, in more than 20 years since the first three-dimensional global simulations, estimates that the Earth will warm by about 3 degrees Celsius for a doubling of carbon dioxide have varied by less than a factor of two? Or that the first simple model calculations by Nobel prize winner Svante Arrhenius in the 1890's, were within that range? Or that the model studies have been reviewed by several prestigious committees of the National Academy of Sciences, with the conclusion: ". . . we have tried but have been unable to find any overlooked or underestimated physical effects that could reduce the currently estimated global warmings to negligible proportions?"

Of course, as Jim Hansen knew only too well, the answer for most of the media, if not for Nicholas Wade, was no, those facts weren't common knowledge among those who write stories on deadline and rely on quotes from scientists they call on the phone to flesh out the parameters of "the debate."

And that delineates one of the major roadblocks to public understanding (and media reporting) of scientific and technological "issues." In a nutshell, there are seldom two sides to an issue in science and technology—there are, instead, many shades of gray and gradations of opinions.

Journalists are taught to balance their reportage of any point of view with a contrasting or opposing point of view. Unfortunately, while legal matters that land in a courtroom crystallize instantly into two opposing camps with opposing points of view, neither life—nor science—is usually that clearcut. Yet, there is scant newspaper space or airtime to adequately present the shades and nuances of an ongoing scientific debate; the tendency is usually to speak to a Stephen Schneider about global warming, then to call a Richard Lindzen for the contrary point of view, and summarize the responses of both as a point-counterpoint proposition.

If you were to do exactly that (call for a summary on global warming), Stephen Schneider would tell you that the GCM's (general circulation models) they run at NCAR, GISS, and elsewhere are "dirty crystal balls," but that all of them agree on eventual warming of between 2° to 5° C due to the added carbon dioxide and other trace greenhouse gases in the atmosphere, and that such warming could have very serious and disruptive effects on human civilization. He would also tell you that whether or not global warming has actually shown up or been proven as an impending disaster, taking certain energy conservation steps now while scientists continue their research is not imprudent.

Richard Lindzen would tell you that the greenhouse effect is certainly beyond question, that added CO_2 unquestionably adds more heat to the atmosphere, and that GCM's are useful, but he would tell you that GCM's are *very* unreliable, and have overstated the expected global warming because modelers such as Schneider and Hansen have failed to adequately take into account the heat sink effect of feedback mechanisms such as the additional clouds produced by extra ocean heating, and the heat absorption ability of the oceans themselves. Since their estimates of the effects of global warming are excessive, in Lindzen's opinion, then there will be little danger for the world if countries such as the United States wait for more research. Immediate unilateral American action, in other words, is premature.

In print, this would be distilled in many cases to Stephen Schneider's calling for immediate action against expected potentially catastrophic global warming, versus Lindzen's saying that the global-warming scare is unfounded because the science isn't certain, and nothing but additional study is needed because we shouldn't take expensive actions until we're absolutely sure.

The average reader skims through such an article and comes away with a joined dichotomy: "We're in deep trouble!" or "We're not in trouble at all."

Whom would *you* believe?

Naturally those who have already made up their minds on environmental issues in general without the need for information or understanding are going to nod at their adopted side and sneer at the other. But for those genuinely interested in finding out the facts in order to make an informed decision, the characterization of such a complex and important issue as a pro and con debate thoroughly garbles and confuses it.

What such a reader will never see in this case is the fact that—despite rather long-standing personal differences between Schneider and Lindzen—there are more areas of agreement than disagreement between the two on the science. Yet in press reports the public sees only "Greenhouse Effect: Fact or Fiction," with Steve Schneider on one side and Dick Lindzen throwing mud balls at him from the opposite side. Such articles almost never show a disagreement as a family debate over matters of degree and shades of gray within the same room of the same scientific house.

Even among philosophical allies, the "balanced reportage" tendencies can create controversies where none really exist. After Jim Hansen's "99 percent confidence" testimony in 1988, Stephen Schneider wasn't the only climatologist pinioned by reporters with questions that left no room for explanation. In his July 8, 1989, guest editorial in *The Washington Post* ("The Greenhouse Climate of Fear"), climatologist Patrick Michaels launched a broadside attack on Jim Hansen's data and his

June 23, 1988, statements. Michaels cited a study on urban warming in which Tom Karl of the National Climate Data Center in Asheville, North Carolina, had questioned the accuracy of the rising temperature records that Hansen (and others) had relied on—records showing a 0.5° C (approximately 1° F) temperature increase over the past one hundred years as perhaps having been biased upward by the heat of cities that during that time were growing up around the weather-sampling sites (such as airports). Hansen had long known of Karl's concerns over urban "heat centers," and had corrected his data to tune out the effects of urban heating on worldwide temperature increases. While he and Karl disagreed slightly on the estimated rate of global temperature increase in the past century (Karl held to 0.4° C, Hansen to 0.5° C), there was no disagreement that there had been a warming. Yet, as soon as Karl's paper was out, he found himself being cited as the "other side of the argument" in disproving the widely quoted finding that the world's temperature had gone up 0.5° C (1° F) in the past one hundred years.

Jim Hansen leaned forward to resume typing his reply to *The New York Times*. The one thing that made him fulminatingly mad—mad enough to interrupt his research to write a "letter to the editor"—was the implication that crying wolf and sounding an alarm was somehow a reprehensible act in and of itself. If the greenhouse issue doesn't deserve a cry or warning, he thought, what does?

He scrolled the cursor to the top of the page and punched in a title: "A Time to Cry Wolf."[8]

> When is the proper time to cry wolf? Must we wait until the prey, in this case the world's environment, is mangled by the wolf's grip?
>
> The danger of crying too soon, which much of the scientific community fears, is that a few cool years may discredit the whole issue. But I believe that decision-makers and the man-in-the-street

can be educated about natural climate variability, and the fact that regional variability presently exceeds greenhouse warming.

A greater danger is to wait too long. The climate system has great inertia, so as yet we have only realized a part of the climate change which will be caused by gases we have already added to the atmosphere. Add to this the inertia of the world's energy, economic, and political systems, which will affect any plans to reduce greenhouse gas emissions. Although I am optimistic that we can still avoid the worst case climate scenarios, the time to cry wolf is here.

CHAPTER

12

Of Senders and Receivers

> *The inhabitants of planet Earth are quietly conducting a gigantic environmental experiment. So vast and so sweeping will be the consequences that, were it brought before any responsible council for approval, it would be firmly rejected. Yet it goes on with little interference from any jurisdiction or nation.*
> Wallace S. Broecker (*Nature*)

Washington, D.C., February 12, 1990

Dr. Robert Watson of NASA leaned forward in the rather rickety swivel chair in his crowded office, ignoring the very real danger that he might fall forward. His eyebrows flared as he warmed to his subject, his words full of expression and feeling as he fielded questions on how to handle the challenges of the greenhouse effect, global warming, and possible climatic change.

"We do know that all of these gases are changing, and they're changing primarily due to human activity. We do know they are chemically active, and they [pose a threat of increasing ozone in the troposphere where we don't want it, and decreasing it in the stratosphere, where we *do* want it], and we also know that these are greenhouse gases that absorb outgoing [infrared] terrestrial radiation, so they will change things. We know that they're changing some of the physics and chemistry of the Earth's atmosphere, and I've often stated I'm more concerned about what we don't know than what we do know."

"It is that unknown that has me perturbed that we should indeed move ahead. Some people hide behind scientific uncertainties as a reason not to do something. I think Steve Schneider is perhaps the most articulate person on this point. He likens environmental protection to buying life insurance or home insurance—you don't expect to be robbed in your home [and] you hope you're not going to drop dead, but all of us, unless you're very poor . . . buy such insurance. And I think it's very similar, having policies that reduce the rate at which we're perturbing the environment is tantamount to buying insurance. They make an awful lot of sense."

But Watson also argues that to dive headlong into major national action without a careful plan based on sound research is equally counterproductive, especially since the maximum impact of a U.S. cutback to a "no-growth of CO_2 emissions" policy would cut less than 5 percent of the world's yearly CO_2 additions.

"I probably agree as much as Jim [Hansen] that global warming is a serious issue, and [that] human-induced global warming is occurring, but I can't look at his data and come up with such a strong conclusion."

Watson stopped for a minute, an uncharacteristic pause, then continued.

"Of course, Sherry [Rowland] for a long period of time was far more convinced about chlorine destroying ozone in large amounts than probably I was, until we came to something like the Antarctic ozone [hole] which sort of hits you around the face. The worrisome thing is, in the global-warming issue or global ozone minus Antarctica, establishing cause and effect is very, very difficult. Jim [Hansen] believes personally it has been established, and he should state that. I would go so far as to say Jim is probably in a minority. Now, he's in a majority in that we're all equally concerned. And so, how do you present this policy argument? Do you have to go as far as Jim did, or can you get the same message across without having to make that final tie?"

Four days before, Dr. D. Allan Bromley, President Bush's science adviser (technically the director of the Office of Science and Technology Policy), had given a carefully prepared statement before another Senate subcommittee in which he stated both the science and the Bush policy: Action by government of the United States to reduce greenhouse emissions must take the form of accelerated research until mitigation efforts can be balanced with economics. In other words, with the scientists themselves differing over how bad the threat really is, American business will not be forced to make sacrifices that might prove to be unnecessary.

"Issues of global change are very important to me and the President. Many people have the impression that the Bush administration is doing less than other countries about global warming and other major environmental problems. That impression is entirely mistaken. If we measure progress in terms of concrete actions, we find that the United States is doing more than any country to understand and address these problems.

"It is imperative that the Bush administration, in cooperation with Congress, develop a well-designed program for research into both the Earth system and the economic policy aspects of potential responses."

The Bush administration would support intensive research (including a near-doubling of research funding), but would take no unilateral action to reduce greenhouse emissions without full international cooperation. George Bush had said as much in person the previous week when he gave a rather distracted speech to the United Nations' sponsored Intergovernmental Panel on Climate Change, calling for "convergence between global environmental policy and global economic policy, a bargain where both perspectives benefit and neither is compromised."

Dr. Bromley underscored the point once again in his prepared statement to the Senate hearing:

"... only strong economies will allow nations to fulfill the obligation of environmental stewardship."

"I think [the President's response] to the situation is as reasonable as you could expect given what his advisers laid on his desk," Jim Hansen said in New York the same day. "You couldn't expect him to do much more than what he's proposing because it does appear to be just a very foggy situation. But I don't think his advisers have served him well. They should be making clear that there is a strong, urgent case for a very aggressive policy of action on things such as improved energy efficiency and development of alternate clean energy sources. Given this situation, what is really important is to make clear to the public what the early signs of the greenhouse effect and global warming are, because I think it's really going to be up to the public to put pressure on the politics to force the issue. There's a lot of [public] concern, and of course, they also hear this scientific debate and have a hard time evaluating the merits of differing points of view, so it's important to be able to communicate to the public."

But who's to do the communicating?

The greenhouse effect–global-warming issue sits at the beginning of the 1990s in the very same position that the chlorine-ozone issue occupied in 1975. There have been initial warnings and "midnight discoveries" that have met with various levels of rejection from within the scientific community and from those to whom any environmentally related alert simply has to be the work of a fringe element of "greenies." Just as public concern and public pressure in 1976–78 (albeit often misguided and based on gross misunderstanding of the science involved, much of it resulting from industry obfuscation) began to force the hand of the U.S. government on the issue of ozone-destroying CFC spray can propellants, so public pressure and concern has pushed President Bush to dedicate funds and political capital to at least *study* the problem, a solution that is by no means unreasonable given—as even Jim Hansen points out—the uncertainties.

But what do we really have here?

This isn't a contest of political wills between those who want to rape, pillage, and plunder this planet and those who, through strict nobility and altruism, want to save it for future generations. This is a responsible debate over what the public policy of the nations of the world should be with respect to CO_2 emissions, global warming, CFC's, and the regrettable human tendency to treat the atmosphere as an eternally forgiving garbage dump.

The problem is not that the Bush administration, Margaret Thatcher's (and now John Major's) government in the United Kingdom, or Helmut Kohl of Germany, François Mitterrand of France, the Japanese or Chinese governments or anyone else who share this gossamer life-support system of atmosphere we all depend on want to blind themselves to future consequences. The problem is dealing with the art of the possible in political and economic terms. Indeed, Dr. Bromley is absolutely right that only stable economies will be able to shoulder the possibly expensive burdens of taking the actions necessary to change the projected outcome of our current worldwide loading of the atmosphere.

But what the good doctor does *not* say is that to get governments—including the United States—to take action, far more than just scientific information is required (no matter how perfect the level of certainty achieved by research). Political will is also needed, and *that* commodity is created only by public interest and public pressure, regardless of the form of government. The ability of scientific findings, facts, and concerns to breed public pressure, however, absolutely *depends* on communication—and that responsibility is a two-way street.

The classic definition of communication is that there has to be a sender and a receiver, or communication cannot occur. With an issue as complex as global warming, *hearing* or *seeing* a report on the subject, however accurate and balanced, is not enough. The receiver must have the rudimentary understanding of science, the scientific process, and the knowledge that very few issues related to science and technology are simple, before there is any hope of understanding. No human on this planet should ever forget that the key to cynical governmental manipulation of entire countries is the failure of the individual citizen to question the message of a Hitler, a Stalin, a Saddam Hussein, or even a Joe McCarthy. Blind acceptance of media-borne information in this increasingly global community—sutured together as we increasingly are by television, telephone, fax, and CNN—means there is a potential of horribly mistaken political actions based on wholly misguided public pressure caused by blind acceptance of simplified versions of the complicated challenges we face as a species, and as a planetary community all riding together on this little blue marble in space. We cannot afford to make major mistakes any longer, because the Earth's population has grown so large that the impact of what we do (or don't do) as a global community can literally change the Earth, and perhaps make it uninhabitable sometime in the future.

The body politic of planet Earth has to have an educated receiver for the increasingly technological and scientific information being transmitted. Otherwise, how on Earth can we direct our political leaders?

Is that a recommendation for everyone to go register in a community college for science courses? Of course not. But it *does* mean that the level of scientific education in the United States, the United Kingdom, Canada, and the rest of the world must be raised several orders of magnitude in priority. In the United States, basic science education is all but a sick joke. We churn out scientific illiterates with every high school graduation, and America is already paying the consequences in terms of reduced brainpower flowing into the research labs charged with solving some of these thorny problems. The problem, however, is international in scope, because there are many nations on the planet who have not faced the reality that cursory introductory courses in scientific and technological disciplines are not enough.

Take genetic engineering, for instance. How many nonscientists can give a reasonably detailed layman's explanation of DNA (deoxyribonucleic acid) and how genes within the DNA structure can be manipulated? Yet we all will be faced with some incredible sociolegal decisions within a matter of decades stemming from great leaps in the ability of genetic engineering. Scientists will be able to do everything from "custom design" human beings in form and appearance, to the much more simple (but incredibly serious) capability of preordaining the sex of a fetus so parents can have a boy or girl on demand—either capability potentially dangerous to humanity. The key to making bad societal decisions on such things lies in the inability of the body politic to understand the science behind such future capabilities. When a population can be swayed with hysterics and disinformation from cynical masters of the art such as Joseph Goebbels of Nazi Germany, the outlook for civilization dims measurably.

The best societal defense against the prospect, however remote, that science and technology could someday become a "black art"—one beyond the basic understanding of the common man and woman—is universal scientific education.[1]

If we have too many citizens who can't understand what the scientists are saying, we also have too many scientists who can't communicate. The universities that prepare such people deserve a few minutes in the line of fire as well for churning out scientists who may be Nobel quality in their scientific knowledge, but who have been shortchanged in their preparation for dealing with the nonscientific community, and left without the basic training to communicate by the written or spoken word in anything other than professional papers and meetings.

As recently as ten years ago a Ph.D. candidate was required to take no more than a single open-book language test to qualify for his or her doctorate, the assumption being that anything in the scientific literature from another country worth reading will be translated into English. Now most science departments have eliminated even that minuscule requirement. Aside from the obvious cultural snobbishness of this institutionalized attitude, it is a symptom of the point of view that scientists-in-training would be wasting their valuable time by taking communications courses in undergraduate or graduate degree programs. Yet scientists are called on every day to communicate verbally, and many of them do an abysmally poor job of it. As Susan Solomon has so correctly pointed out, a scientist has to make his or her own decision whether to be an advocate, an activist, or a quiet researcher. But even the quietest, newest, most intimidated, unpublished, and insecure postdoc in the lab should be able to stand up and deliver a clear and concise account of himself and his research even within the confines of that particular branch of science. The scientific educational house, in other words, is not in order on this point.

Scientists such as Steve Schneider and Carl Sagan are rarities. They irritate their brethren by actually *enjoying* the process of communication with the public and the media, and they occasionally make mistakes or inadvertently fan the flames of controversy real or imagined. Scientists such as James Hansen and Sherry Rowland are also rarities. They may not enjoy the process as much, but they have the courage to state their conclusions—right or wrong—in public and for the greater good, helping to bring the inner workings of the disciplined scientific minds engaged in complex debates often over shades of gray to the attention and comprehension of the average layman. And that, especially for a free society, is an invaluable function.

Science is not a religion with a priesthood expected to keep secrets and traditions and innermost dogmatic practices to themselves as the province of the clergy. Science is the human quest for knowledge pressed forward in an organized, disciplined manner with its own set of rules. But if scientists cannot communicate, then the media cannot translate, and the people cannot understand, and knowledge gained is then effectively lost (or at least sequestered to the exclusive use of the "priesthood").

But this is not to devalue or decry the bulk of the scientific community who "daily toil with the data" (as Patrick Michaels put it in his article "A Climate of Fear") and who choose not to belly up to the public forum. Nevertheless, there is a concurrent responsibility for such men and women that runs parallel with their task of keeping scientific debate alive and healthy through constant questioning and testing of the scientific process, and that is simply this: Attacks on the veracity of a fellow scientist should be made only on the basis of a legitimate need to peer review and question the science, not for purposes of pique, revenge, jealousy, or indirect expression of disgust with fellow scientists who would speak to the public and the media directly rather than through published papers.

The public does not read—nor could they understand—published scientific papers. Most Americans have a hard time understanding the articles in *Scientific American*. Therefore the Sagans and the Schneiders and the Rowlands and the Hansens are very necessary to the process of communication, as are the Richard Lindzens and Tom Karls (when properly quoted) in balancing the equation.

In an address to the American Association for the Advancement of Science on February 19, 1990, Steve Schneider spoke of the polarization of scientific issues by the public impression of catfights among scientists carried out in the media:

> Not all knowledgeable scientists are in agreement as to the probability that such changes [as global warming-caused sea level rise and climate shifts] will occur. In fact, if one has followed the very

noisy, often polemical debate in the media recently, one might get the (I believe false) impression that there are but two radically opposed schools of thought about global warming: 1. that climatic changes will be so severe, so sudden and so certain that major species extinction events will intensify, sea-level rise will create tens of millions of environmental refugees, millions to perhaps billions of people will starve and devastated ecosystems are a virtual certainty or, alternatively, 2. there is nothing but uncertainty about global warming, no evidence that the twentieth century has done what the modelers have predicted, and the people arguing for change are just "environmental extremists"—thus there is no need for any management response to an event that is improbable and in no case should any such response interfere with the "free market" and bankrupt the nation.

Schneider referred to a vicious cover article in *Forbes* magazine (December 25, 1989) entitled "The Global Warming Panic" as an example of the alarmed counterattack some elements of the business community have felt it necessary to mount.[2]

Unfortunately, while such a highly charged and polarized debate makes entertaining opinion page reading or viewing for the ratings-dominated media, it provides a very poor description of the reality of the actual scientific community. In my opinion, the "end of the world" or "nothing to worry about" [scenarios] are the two *least likely* cases, with almost any scenario in between having a higher probability of occurrence.

* * *

What the world needs now—as the preamble to the 1960s song goes—is far better *communication* among scientists, policymakers, the media, and the public. In Bob Watson's words:

"What you need is a spectrum of people that go from a hardworking scientist or field scientist all the way through to someone to translate [it] from hard science, which has to be completely intellectually defensible to the policymaker. I think it needs scientists like Sherry Rowland, who as an individual has recognized an issue [and] had the courage of his convictions, [and] did some really good science in the meanwhile, and has tried to hammer home to governments that there's a fundamental problem. It also needs, though, a group of scientists that have not stated any personal convictions to policymakers about what they would do if they could ... a group of scientists that indeed may well have some personal views, but fundamentally have only got their eyes on a scientific assessment absolutely divorced from the politics or policy considerations."

From Jim Hansen's point of view:

"I realized [in testifying in 1986] how hard it is to make clear to even a congressman—and there are some very smart guys down there in D.C.—why [for example] it is important that the temperature is going up globally. You have to think of simple examples, such as how many days you'll have over one hundred degrees [if global warming kicks in] in order to make the story clear. And some scientists don't like that."

And Steve Schneider:

"The biggest problem with scientists is that they're uncomfortable as a culture with intuitive probability, and they're uncomfortable with the policy process that demands that before their probabilities become objective, or the evidence becomes overwhelming, you simply don't speak up. In my case I learned this in 1972, that I was going to have a lifetime of trouble with a lot of people, and yet it was going to be over their misunderstanding partly caused by the press extrapolating what you say, and partly caused by their own ignorance of the policy process.

"All science can do is provide probabilities and consequences. How we react to those probabilities and consequences is in the public realm, and for scientists to decide how to interpret probabilities and consequences is the height of elitism and is nondemocratic. It [says] that democracy isn't working worth a damn."

The system does work, but perhaps not fast enough. The public hears, but perhaps not clearly enough, and with insufficient understanding of the nuances. The political world listens, but expects the scientists to make their voices heard above the opposing clamor of scientific debate, which is mistaken as invalidating urgency or seriousness of the problem short of 100 percent agreement.

We all must learn that science is never sure of anything. Thus, waiting for perfect agreement and "final findings" is like Waiting for Godot—it will never happen. A nation that waits for perfect certainty among its scientists will be confronted by Antarctic ozone holes—uncontrollable "events" that drive desperate reactions and deny the society the chance to minimize and avoid emergencies through advanced planning.

We also have to learn to understand the *medium* as well as the message it transmits. Radio and television simply do not have the luxury of time to explain all the nuances and balances that are needed to make rational public decisions. *USA Today*, for instance, is a marvelous tool for drawing the nation together in terms of access to local and regional news from one side of the country to another, but neither it nor most local newspapers can possibly present all the information needed to fully understand a difficult issue over which reasonable people can differ. When it comes to complicated science and technological controversies, we have to learn never to accept the headline as the complete story. Seldom is a scientific debate ever a clear-cut case of this-side-versus-that-side.

In the ozone issue, as in the global-warming issue, we have serious, responsible people in science, government, the media, and the economy all trying to do the right thing at the right time in accordance with the interests of us all. Some are better at it than others, some more self-interested than others, and in truth there are few if any villains. However, there is at least one truth beyond debate: A representative democracy whose members cannot communicate clearly and in depth on complicated scientific and technological issues has a lesser ability to apply human intellect to avoid societal disasters.

That goes beyond theory.

That is an axiom.

APPENDIX I

Update on The Main Players in this Story

As of the spring of 1991 . . .

Dr. Sherwood F. (Sherry) Rowland continues to serve as a full professor of chemistry at the University of California at Irvine, and continues to fill the role of elder statesman in the ozone-chlorofluorocarbon issue. Contrary to the professional difficulties he encountered previously for his advocacy for CFC limits, he is now one of the most sought-after figures in atmospheric chemistry worldwide. In October 1987, Dr. Rowland received the prestigious Charles A. Dana Award for pioneering achievements in health, and in 1989 the Science and Technology Foundation of Japan awarded him the 1989 Japan Prize, one of the world's top honors in scientific research, which also carries a $400,000 monetary award.

Dr. M. J. Molina, now Professor, Department of Earth, Atmospheric and Planetary Sciences and Department of Chemistry at MIT, continues his energetic and pace-setting research in areas of atmospheric chemistry. Dr. Molina has received the Esselen Award of the American Chemical Society (1987) and most recently the Newcomb-Cleveland prize of the American Association for the Advancement of Science for his elegant 1987 paper in *Science* describing his work on the Antarctic Ozone Hole chemistry (postulation of the ClOOCl "dimer" and the methods of chemical denitrification and dehydration in PSC's). He continues to be a major player on the world stage of understanding the chemical processes threatening our atmosphere.

Dr. Susan Solomon continues to be one of the key researchers at the Aeronomy Laboratory of the National Oceanic and Atmospheric Administration's Boulder facility, and one of the world's pacesetters in atmospheric chemistry. Despite her professional modesty, her contributions to this all-important field are difficult to overemphasize.

Dr. Robert Watson, Director of the Office of Space Science and Applications of the Earth Science and Applications Division of NASA in Washington, D.C., has accelerated his worldwide stewardship of a host of research progams involving global atmospheric change and ozone depletion. His schedule would kill the average twenty-year-old.

Dr. Joe Farman retired in 1990 after a distinguished career spanning the better part of four decades, most of it with the British Antarctic Survey. He lives near Cambridge, England.

Dr. Stephen Schneider continues his research and climate modeling as well as his international advocacy of climatological responsibility as a leading climatologist with the National Center for Atmospheric Research in Boulder, Colorado. As a speaker, communicator, and author, he spends much of his professional time focusing the nations of the world on the hazards of inadvertent climate modification.

Dr. James Hansen, Director of the Goddard Institute for Space Studies of NASA in New York City, continues to pursue his team's research projects in advanced climate modeling while remaining accessible to the media and the public. His office remains above Tom's Restaurant in upper Manhattan near Columbia University.

Dr. James Anderson, who continues as a professor at Harvard, is pushing the pace-setting research of his various team members into new and innovative methods of upper stratospheric sampling, including the use of unmanned drone aircraft that can stay aloft for days at a time and fly literally to the poles and back gathering invaluable information on the chemical and dynamical realities of our changing atmosphere. He has managed at least one vacation in Idaho during the past year.

Dr. Adrian Tuck, who is no longer hirsute (possessed of a mustache, whether walrus *or* Mexican bandit), remains deeply involved in the issues of Arctic and Antarctic ozone destruction from his office at NOAA's Aeronomy Lab in Boulder.

APPENDIX II

Acknowledgments

The scientific method involves the constant testing and evaluation of ideas and theories constructed to explain the universe in which we find ourselves. For the writer-journalist-researcher who would penetrate a scientific discipline without benefit of a Ph.D. in that field (with the intent and purpose of explaining it to his readership), the process is the same. The initial ideas and tentative conclusions of the writer must be tested brutally and constantly against the realities of the science *as seen by those within that science*!

It would be virtually impossibl , then, to write a quality book of this scope and caliber without a major contribution of time and cooperation by the professionals in the various sub-disciplines of atmospheric science. While the ultimate conclusions and the thesis of this work are entirely mine (and while the responsibility for ultimate accuracy of the facts and conclusions is also mine), a lion's share of the credit for whatever effectiveness *What Goes Up* achieves in terms of increasing international understanding of these vital issues belongs to a great number of people in the scientific community who have been unfailingly patient and generous with their time, their reference material and papers, and the benefit of their intellect. Chief among them are the following individuals:

Appendix II [287]

Dr. Sherwood F. Rowland—Professor of Chemistry, University of California at Irvine

Sherry Rowland had been described to me as friendly and patient long before I met him in Snowmass, Colorado, during the weeklong conference in 1988, and indeed, those qualities were instantly apparent when this writer, whom he did not know, approached him in the lobby of the conference center between sessions with what (at that stage in the research) were extremely basic questions. With professorial ease and interest, out came his pen, and he began to diagram for me on the pages of my notebook chemical reactions that a first-year chemistry student would find rather pedestrian. As my questions increased in complexity over the next few days (and, I hope, in sophistication over the following two years), Professor Rowland's patience never flagged, and his profound interest in accuracy never waned. When I sent a majority of this manuscript to him for comment and technical correction (as I did with a circle of senior scientists), I did expect that he would, as promised, spend some time looking over the book. I was not prepared, however, for the comprehensive and detailed pages of handwritten notes I received back from him within a few weeks—voluminous explanations of the events I had described as well as notations refining my understanding of the chemical truths and discoveries covered—pages which must have taken many hours to prepare, and which will retain an honored place in my library of research material. While I must emphasize that the words and conclusions in this book are entirely mine and must not be imputed to Dr. Rowland or any other party except where quoted, his gracious assistance has been quite pivotal, and entirely definitive of a true professor.

Dr. Mario J. Molina—Professor, Department of Earth, Atmospheric and Planetary Sciences and Department of Chemistry, MIT.

Mario Molina's willingness to spend time with me at Snowmass, explaining yet again the key struggles in the ozone equation over the prior fourteen years despite the pressures of the moment, was followed by unflagging patience in taking my phone calls, and providing careful analysis and markup of the many pages of this manuscript I asked him to review. His contributions to the accuracy of this work are many, and are greatly appreciated.

Dr. Susan Solomon—Atmospheric Chemist, Aeronomy Laboratory, National Oceanic and Atmospheric Administration, Boulder, Colorado

My first contact with Susan Solomon came in the form of my phone call out of the blue to her Boulder office. In my roles as an airline safety analyst, and earthquake safety advocate, I, too, deal with constant, unsolicited calls for information and comment from the media and fellow journalists day in and day out, and I try very hard always to be receptive and helpful whether I know the caller or not. Susan, however, set new standards for me in terms of graciousness and receptivity. Her willingness to explain yet again what she had probably detailed to reporters a thousand times over about her subject—and her willingness over the following years to field my calls, check the manuscript, and meet with me several times in Boulder—have been invaluable assets to this work.

Dr. James Hansen—Director, Goddard Institute for Space Studies, NASA, New York City

I had never met or talked with Jim Hansen before the day I walked into his office in upper Manhattan to keep an appointment made the week before from Seattle. Because his life had been altered rather drastically the year before by the intense glare of publicity, I found him to be very guarded and reserved in his direct statements, but very willing and able to provide me with voluminous copies of papers, articles, statements, and other material which are really vital to researching such a book as this. In later phone calls, however, and in his review of the chapters I sent to him for comment, he has taken as much care and time as Sherry Rowland in fairly and carefully explaining anything and everything I wanted to know. The thread of intense personal integrity and intellectual honestly which I have come to admire so much in such individuals as Sherry Rowland, Mario Molina, Susan Solomon, and their fellows in this discipline, is exemplified by Jim Hansen.

Dr. James Anderson—Professor, Harvard University

As with his fellow world-class scientists, Jim Anderson—from talking with me in Snowmass through fielding my phone calls and manuscript review requests—has been unflaggingly helpful and friendly. His vacation home in Idaho is not too distant from my own home in the verdant forests of Western Washington, and I probably felt more acutely than most the sacrifice of his having to leave such a place to shoulder the task Bob Watson handed him prior to the Punta Arenas expedition.

Dr. Robert Watson—Director, Office of Space Sciences and Applications Division, NASA, Washington, D.C.

"You haven't met Bob Watson yet?" asked the senior scientist of me at Snowmass. "Boy, are you in for a treat!"

I'm not sure I believed that at the time, but it turned out to be an understatement. Through the substantial assistance of Dr. Watson's associate Flo Ormund, we finally arranged to pin him down for a Washington interview which was supposed to run only an hour on a crisp, cold January day. Three hours of nonstop information and impassioned discussion later, I wobbled out of his office carrying tapes of the interview which would literally fill a binder on my shelf when eventually transcribed. Dr. Watson is indeed a national asset, and this book has been greatly assisted by his participation.

Dr. Adrian Tuck—Atmospheric Scientist, Aeronomy Laboratory, National Oceanic and Atmospheric Administration, Boulder, Colorado

No matter how dedicated the journalist, and no matter how important the book being prepared, having a writer camp in your office with a tape recorder and notebook for several hours simply wipes out a substantial portion of the working day. This is especially true when the request for such an interview comes at the last moment, by phone, from a Denny's a mile distant. I very much appreciate Adrian Tuck's receptiveness to that particular intrusion, and his friendliness and helpfulness at Snowmass and through several phone calls, as well as his help with review of portions of the manuscript. And I particularly appreciate his keen sense of humor and taste in classical music.

Dr. Stephen H. Schneider—Head of the Interdisciplinary Climate systems group of the Climate and Global Dynamics Division, National Center for Atmospheric Reserch (NCAR), Boulder, Colorado

The day I invited Steve Schneider to lunch following an interview, he ended up with little more than a snack from the cafeteria at NCAR, so intent was he on answering my questions and assisting my understanding of the global warming issues in the time he had between other commitments. I still owe him a decent lunch. Dr.

Appendix II

Schneider has reviewed pages of this manuscript on airplanes and fielded telephone calls on the run to help me, and I greatly acknowledge and appreciate his help and his personal insight, and especially his candor.

Dr. Joe Farman—Scientist, British Antarctic Survey (Retired)

For the time that Dr. Farman spent with me at Snowmass discussing his pivotal contributions to the discovery of the Antarctic Ozone hole, and for the time he has spent on the phone with me prior to his retirement in 1990, I am very grateful.

In addition, the following people were of great assistance with the research writing, checking, and logistics behind *What Goes Up*:

Dr. Arthur Aikin—NASA
Dr. Daniel Albritton—NOAA
Dr. John Austin—University of Washington
Dr. Ralph J. Cicerone—NCAR
Dr. Estelle Condon—NASA Ames
Ms. Liz Cook—Friends of the Earth
Dr. Paul Crutzen—Max-Planck Institute for Chemistry, Mainz, Germany
Dr. Diane Fisher—Environmental Defense Fund
Dr. Neil Harris—University of California at Irvine
Dr. Dennis Hartman—University of Washington
Mr. David Harwood—Staff Member, Sen. Tim Wirth-Colorado
Dr. Donald Heath—NASA
Dr. D. J. Hofmann—University of Wyoming
Dr. Geoff Jenkins—UK Department of the Environment
Dr. W. D. Komhyr—NOAA
Dr. Arlin J. Krueger—NASA
Dr. Richard Lindsen—MIT
Dr. John Lynch—National Science Foundation
Dr. Michael McIntyre—University of Cambridge
Ms. Flo Ormund—NASA, Washington, D.C.
Ms. Sharon Roan—Author, Writer
Dr. James Rosen—University of Wyoming
Dr. Mark Schoeberl—NASA
Dr. Paul C. Simon—Institut d'Aeronomie Spatiale, Brussels

Dr. Richard Stolarski—NASA
Dr. N. Sundararaman—World Meteorological Organization, Geneva
Dr. Steven Wofsy—Harvard

I would also like to thank Steve Schneider's secretary, Leigh, and Jim Hansen's secretary, Carolyn Paurowski, for their kind assistance in arranging interviews and sending materials.

APPENDIX III

References

The lessons of the ozone wars and the challenge of the global-warming–greenhouse debates are merely an introduction to the serious questions of atmospheric modification that face all humans on earth. In these issues, as well as with the issue of garbled communication among scientists and the lay community, there is a rather profound truth: The more you know, the be ter prepared you are to participate in designing and implemen ing the solutions. To that end, the source material available to you in bookstores, libraries, from the federal government, and through universities, is voluminous. Magazine articles, books, detailed scientific papers, and conference publications, among other sources, can take you anywhere from a slightly deeper perusal of these subjects to a professional scientific level of familiarity.

To begin with, there are two excellent books of current vintage which can take you into the subjects of global ozone loss and global warming respectively. *Ozone Crisis, The Fifteen Year Evolution of a Sudden Global Emergency,* by Sharon L. Roan (John Wiley and Sons, New York, 1989), is a comprehensive and authoritative examination of the ozone wars in greater depth and technical detail than could be presented in this work. It, too, is aimed at the general reader and thoroughly understandable, and I highly recommend it to you.

I would also call your attention to *Global Warming, Are We Entering the Greenhouse Century?*, by Dr. Stephen H. Schneider of NCAR (Sierra Club Books, San Francisco, 1989), which takes you in equal depth into the greenhouse–global-warming equation, as well as into Steve Schneider's experiences in trying to be a communicator and catalyst between the scientific and lay worlds.

For more background detail on the 1970s' intramural combat over ozone, the 1978 book *The Ozone War*, by Harold Schiff and Lydia Dotto (New York, Doubleday, 1978), is required reading. The book is available through most major library systems.

And, an additional current book which provides an entertaining romp through both subjects is *The Next One Hundred Years, Shaping the Fate of Our Living Earth*, by Jonathan Weiner (Bantam Books, New York, 1990).

For the more dedicated student of these subjects, I would recommend several evenings in a large library poring through the citations of current articles in popular and professional magazines. These are very current, changeable, and vibrant subjects which are quite properly drawing a great amount of attention (and "ink") from journalists worldwide, and a comprehensive tour through the resulting articles can be very revealing, not only in terms of the subject, but also in terms of the various levels of media understanding/misunderstanding and oversimplification. It is not at all difficult to prove yet again the thesis of this book (that we aren't communicating clearly or adequately) with such research.

The next level into either global warming or ozone (or indeed atmospheric science in any of its permutations) requires a careful trek through current publications of NOAA, NASA, and associated professional papers published in the leading scientific journals such as *Science* and *Nature* and other such offerings. The various hearing records of House and Senate subcommittees concerned with atmospheric issues should also be consulted. Specifically, in regard to ozone, I would encourage a thorough reading of the latest report of the International Ozone Trends Panel for 1988 (issued in early 1991), Volumes 1 and 2. For information on obtaining copies of this document, write the World Meteorological Organization, 41, Avenue Giuseppe Motta, Case Postale No. 5, CH-1211, Geneva 20, Switzerland. Also current and vital is the NASA Reference Publication 1242, "Present State of Knowledge of the Upper Atmosphere 1990: An Assessment Report (to the Congress)."

I would also encourage the serious student of these subjects who is not already associated with a major university's atmospheric science, aeronomy, atmospheric chemistry (or other associated) program to make contact with the faculty in such a department for ad hoc guidance in further research and understanding.

And finally, whatever your level of interest, keep in mind the details of the inside stories in this book—stories which demonstrate that behind almost every "scientific controversy" there is a far more detailed and far less polarized truth. If we refuse to take oversimplified reports at face value, our collective ability to properly assess threats discovered by the scientific world and make the appropriate societal/political decisions will be greatly enhanced.

Additional Selected References Utilized in This Work

It would be impractical and of little use to most readers to list every scientific paper, magazine, or newspaper article I've read or used, but for the serious reseracher who would like to track my sources in great detail, a short list might be appropriate. Although several hundred hours of nose-to-nose interviews with scientists and professionals in these field (many of those interviews recorded) have formed one of the basic underpinnings of research for this work, the written record is also vital. The following is a selected list of those materials.

NEWSPAPER ARTICLES

"After Us, the Deluge," Vivian Gornitz, *Newsday*, August 1, 1989, p. 50.
"Global Threat—New Culprit Is Indicted in Greenhouse Effect:

Rising Methane Level," Jerry E. Bishop, *The Wall Street Journal*, October 24, 1988, p. 1.

"Global Warming Flap Gets Hotter," Joseph B. Verrengia, *Rocky Mountain News*, February 20, 1990, Pg. 8M.

"Governments Start Preparing for Global Warming Disasters," William Stevens, *The New York Times*, November 14, 1989, p. C1.

"Greenhouse Study Targets Area's Future," Len Maniace, *The Journal News* (Gannett), September 11, 1989.

"I'm Not Being an Alarmist About the Greenhouse Effect," James Hansen, Editorial to *The Washington Post*, February 11, 1989, p. A23.

"Iowan James Hansen Is Prophet of Global Warming," David Goeller, *The Cedar Rapids Gazette*, November 26, 1989, p. 9A.

"Loads of Media Coverage," Editorial, *Detroit News*, November 22, 1989.

"Major Greenhouse Impact Is Unavoidable, Experts Say," Philip Shabecoff, *The New York Times*, July 19, 1988, p. C1.

"News Plays Fast and Loose with the Facts," Stephen H. Schneider, *Detroit News*, December 5, 1989.

"Thou Shalt Not Mess with the Global Environment," *The New York Times*, January 11, 1989, Letters.

"Warming of Earth Is Slowed, Scientist Says," Associated Press Dispatch, *The New York Times*, November 30, 1989, p. A29.

MAGAZINE ARTICLES

"Annals of Chemistry (The Ozone Layer)," Paul Brodeur, *The New Yorker*, June 9, 1989, p. 70.

"A Warming World," Anthony Ramirez, *Fortune* magazine, July 4, 1988, pp. 102–107.

"Bringing NASA Down to Earth," Eliot Marshall, *Science,* June 16, 1989, Vol. 244, pp. 1248–1251.

"Endless Summer: Living with the Greenhouse Effect," Andrew C. Revkin, *Discover* magazine, October, 1988, p. 50.

"Fear of Flooding: Global Warming Could Threaten Low-Lying Asia-Pacific Countries," Michael Malik, *Far Eastern Economic Review* (Hong Kong), December 22, 1988, Vol. 142, Issue 51, pp. 20–21.

"Feeling the Heat on the Greenhouse," *Newsweek*, May 22, 1989, p. 79.

"Greenhouse Skeptic Out in the Cold," Richard A. Kerr, *Science,* December 1, 1989, Vol. 246, pp. 1118–1119.

"Has Stratospheric Ozone Started to Disappear?," Richard A. Kerr, *Science,* Vol. 237, July 10, 1987, pp. 131–132.

"Is It All Just Hot Air?," *Newsweek,* November 20, 1989, p. 84.

"Keep Your Cool," T. A. Heppeneimer, *Reason* magazine, January, 1990, pp. 22–27.

"Layers of Complexity in Ozone Hole," *Science News,* March 14, 1987, pp. 131.

"Living in the Greenhouse," *The Economist,* March 11, 1989, pp. 87–90.

"Man-made Greenhouse: Has it Arrived?," *Geotimes,* August, 1989, p. 12.

"More Clues to the Mysterious Ozone Hole," *Science News,* Vol. 132, September 19, 1987, pp. 182.

"Rethinking the Greenhouse: The Backlash Against Global Warming Has Begun," *The Economist,* December 16, 1989, p. 14.

"Scientist Says Greenhouse Warming Is Here, *Science News,* July 2, 1988, p. 4.

"The Changing Climate," Stephen H. Schneider, *Scientific American,* September, 1989, pp. 70–79.

"The Endless Simmer," Dick Russell, *In These Times,* January 11, 1989, p. 10.

"The Global Warming Panic," W. T. Brookes, *Forbes,* December 25, 1989, pp. 96–102.

"The Heat Is On," Christopher Flavin, *World Watch,* December, 1988, p. 10–20.

"The Struggle to Save Our Planet," *Discover* magazine, Special Edition, April, 1990.

"The Whole Earth Agenda," *U.S. News and World Report,* December 25, 1989, p. 50.

"Winds, Pollutants Drive Ozone Hole," *Science,* Vol. 238, October 9, 1987, pp. 156–158.

PROFESSIONAL PAPERS

Ahladas, John A., "Global Warming, Fact or Science Fiction," Address delivered before the Richmond Jaycees, Richmond, VA, February 7, 1989.

Bromley, D. Allen, Statement before the Subcommittee on VA/ HUD and Independent Agencies of the Senate Committee on Appropriations, February 8, 1990.

Farman, J. C.; Gardiner, B. G.; Shanklin, J. D.; "Large Losses of Total Ozone in Antarctica Reveal Seasonal ClO NO Interaction," *Nature*, Vol. 315, May 16, 1985.

Garcia, Rolando R.; Solomon, Susan, "A Possible Relationship Between Interannual Variability in Antarctic Ozone and the Quasi-Biennial Oscillation," *Geophysical Research Letters*, Vol. 14, No. 8, pp. 848–851, August, 1987.

Hansen, James E., "The Greenhouse, The White House, and Our House," Address given to International Platform Association, Washington, D.C., August 3, 1989.

Hansen, James E., "A Time to Cry Wolf," reply to editor, submitted to *The New York Times*, July, 1989.

Hansen, J.; Fung, I.; Lacis, A.; Lebedeff, S.; Rind, D.; Ruedy, R.; Russell, G.; Stone, P.; "Global Climate Changes as Forecast by Goddard Institute for Space Studies Three-Dimensional Model," *Journal of Geophysical Research*, Vol. 93, No. D8, pp. 9341–9364, August 20, 1988.

Hansen, J.; Fung, I.; Lacis, A.; Lebedeff, S.; Rind, D.; Ruedy, R.; Russell, G.; Stone, P.; "Prediction of Near-Term Climate Evolution: What Can We Tell Decision-Makers Now?," from Proceedings of the First Northern American Conference on Preparing for Climate Changes, October 27, 1989.

Hofmann, David J.; Solomon, Susan, "Ozone Destruction Through Heterogeneous Chemistry Following Eruption of El Chichón," *Journal of Geophysical Research*, Vol. 94, No. D4, pp. 5029–5041, April 20, 1989.

Karl, T. R., and Jones, P. D., "Urban Bias in Area-Averaged Surface Air Temperature Trends," *Bulletin of the American Meteorological Society*, 70:265–270, 1989.

Lindzen, R., "Some Coolness Concerning Global Warming," *Bulletin of the American Meteorological Society;* 1990.

Mount, G. H.; Sanders, R. W.; Schmeltekopf, A. L.; Solomon, S., "Visible Spectroscopy at McMurdo Station, Antarctica," *Journal of Geophysical Research*, Vol. 92, No. D7, pp. 8320–8328, July 20, 1987.

National Academy of Sciences, *Current Issues in Atmospheric Change*, National Academy Press, Washington, D.C., 1987.

Poole, Lamont R.; Solomon, Susan; McCormick, M. Patrick; Pitts, Michael C.; "The Interannual Variability of Polar Stratospheric Clouds and Related Parameters in Antarctica During September and October," *Geophysical Research Letters*, Vol. 16, No. 10, pp. 1157–1160, October, 1989.

Rowland, F. Sherwood; Blake, Donald R.; Stratospheric Feedback from Continued Increases in Tropospheric Methane," Paper presented to Snowmass conference, May, 1988.

Schneider, Stephen H., "The Global Warming Debate: Science or Politics?," Address to the American Association for the Advancement of Science, February 16, 1990.

Solomon, P.; Barrett, J.; Jaramillo, M.; deZafra, R.; Parrish, A.; Emmons, L.; "Daytime Altitude Profile of ClO over McMurdo in September, 1987, and the Mechanism for Ozone Depletion," Paper, presented to Snowmass conference, May 10, 1988.

Solomon, Susan, "The Mystery of the Antarctic Ozone 'Hole'," *Reviews of Geophysics*, Vol. 26, No. 1, pp. 131–148, February, 1988.

Solomon, Susan; Garcia, Rolando, R.; Rowland, Sherwood F.; Wuebbles, Don J.; "On Depletion of Antarctic Ozone," *Nature*, Vol. 321, pp. 755–758, June 19, 1986.

Solomon, Susan; Schmeltekopf, Arthur L.; Sanders, Ryan W., "On the Interpretation of Zenith Sky Absorption Measurement," *Journal of Geophysical Research*, Vol. 92, No. D7, pp. 8311–8319, July 20, 1987.

NOTES

PROLOGUE

1. In Antarctica, the process is as follows:

$$(Cl + O_3 \rightarrow ClO + O_2)\ 2X$$
$$ClO + ClO \rightarrow {}^m Cl_2O_2$$
$$Cl_2O_2 + h\nu \rightarrow Cl + ClO_2$$
$$ClO_2 \rightarrow {}^m Cl + O_2$$

Net Result: $2O_3 \rightarrow 3O_2$

The basic Rowland-Molina cycle, however, is:

$$Cl + O_3 \rightarrow ClO + O_2$$
$$ClO + O \rightarrow Cl + O_2$$

Net Result: $O + O_3 \rightarrow 2O_2$
 (This process is negligible in Antartica)

CHAPTER 1

1. The spectrophotometers used by the British Antarctic Survey measured ozone and other trace gases by analyzing the spectrum from sunlight passing through the air. The units were first developed in 1931 by a British scientist, Sir G. M. B. Dobson, and the units that marked the graduations of the measurements were named Dobson Units after him. Dobson had designed this instrument to help meteorologists study air currents in the atmosphere and measure the amount of ultraviolet light that reaches the earth (by measuring the amount of ozone, which captures certain wavelengths of UV light). The greater the levels of ozone measured, the lower the amount of UV light reaching the surface. The instruments, for all the brilliance of their design, were somewhat antiquated by the 1980s, and the specific ones used by Joe Farman's group had been in use for some time as well, and were difficult to calibrate—

[300] *Notes*

a procedure that required many sequential readings through a large number of sightings at various angles to the vertical.

2. A number that varies exponentially is raised or lowered by a "power." For instance, raising the number 10 to the number 100 and then to the number 1,000 is an exponential progression ($10^1 = 10, 10^2 = 100, 10^3 = 1,000$). By contrast, a linear progression would be something on the order of: 10, 11, 12, 13, 14, 15, 16, 17, 18, 19, 20, etc., or 10, 20, 30, 40, 50.

CHAPTER 2

1. High-flying jet engines leave behind NO (nitric oxide) and NO_2 (nitrogen dioxide).

2. Mario Molina had received his early education in Switzerland, graduating from the University of Mexico with a bachelor's degree in chemical engineering, then attending both the University of Freiburg in Germany, and the Sorbonne in France before electing to study for his Ph.D. under Dr. George Pimentel at the University of California at Irvine, where he joined Sherry Rowland's research group. Molina had just turned thirty at the time.

3. CFC's are not lighter than air, they are heavier, which seems a contradiction. The molecular weight for CCl_3F, for instance, is 137.5 versus a molecular weight of 29.0 for "air." As Professor Rowland explains it: "The mixing of molecules occurs in the atmosphere in large air masses, taking heavy and light at the same rate. When eventually caught in an updraft, they all go up together. The motion is random, up, down, down, up, up, etc., and eventually a string of 'ups' [propels] the CFC's into the stratosphere."

4. Rowland and Molina's findings at that point are summarized as follows: "No tropospheric sinks; photolysis in the stratosphere, but only after 40–80 years for CCl_3F, and 75–150 years for CCl_2F_2; and photolysis released Cl atoms."

5. Molina actually went to the lab to repeat and extend the measurements of UV-C cross sections for CFC's, since data developed by Du Pont had been incomplete. Cross sections provide the quantitative knowledge needed about the rates at which molecules decompose (photodissociate). Both he and Rowland assumed that if a molecule intercepted UV-C, the molecule would fall apart in *every* case. A year later they proved that assumption.

CFC's react with the class of ultraviolet wavelengths roughly classified as UV-C. UV-C is generally defined as wavelengths of ultraviolet below 230nm. UV-B, the wavelengths of UV so damaging to carbon-based DNA on Earth's surface, is roughly defined as 280–320 nm, while UV-A is in the range of 320–400 nm. The service provided by O_3 (ozone) molecules that has permitted life as we know it to develop and flourish on Earth is the continuous absorption of wavelengths of UV-B in the 280–320 nm range. That absorption of energy is translated through the breakage of the molecular bonds into heat energy at stratospheric heights, which gives rise to the inversion heat layer in the stratosphere that in itself defines the vertical delineation of the stratosphere.

6. The stratospheric answer: Chlorine's reaction with ozone is about a thousand times more likely than its reaction potential with anything else. So chlorine reacts with ozone to give ClO—chlorine monoxide.

7. The chlorine atom is a free radical too, with seventeen electrons. ClO has twenty-five electrons.

8. Sherry Rowland explains the process this way: "When chlorine is liberated from CFC's, it comes off as *atomic* chlorine, Cl. The reaction with ozone is the first step in the chain, giving ClO. The second step is ClO + O, completing the first cycle of the chain reaction."

9. Other scientists had done similar work, but they were in the stratospheric research arena, and unknown to Rowland and Molina. At the same time that Mario Molina was working with these figures, the same catalytic chain was being debated by others in relation to the stratosphere. No one, however, had found a source for massive amounts of chlorine. There was nothing as yet in the literature, though Harold Johnson had received a preprint of a paper on the subject from Cicerone and Stolarski.

10. When Molina and Rowland began calculating the atmospheric lifetimes of CFC's in the 40–150 year range, they routinely calculated how much CFC would be in the atmosphere after 300–400 years of continuous emission. Those numbers were much higher than the known 1973 concentrations, so they almost immediately escalated to and focused on the future problem.

11. Two months before Mario Molina's figures reached their startling values regarding ozone destruction, there had been a scientific meeting in Kyoto, Japan, during which Richard Stolarski and Ralph Cicerone proposed that chlorine chemistry might pose some sort of threat to the ozone layer. Teams at Harvard and the University of Michigan were preparing papers on the subject by December of 1973, but neither was considering chlorofluorocarbons as a source of chlorine in the stratosphere. In her book *Ozone Crisis,* Sharon Roan details the original involvement of two scientists—neither of them chemists—who would become a major part of the ozone controversy for the next decade, Cicerone and Stolarski, both of whom determined in 1973 that the space shuttle's exhaust would produce chlorine that would attack and deplete as much as a tenth of a percent of the world's supply. The official reaction was "So what?" but unofficially NASA was frightened enough to suggest that such findings really didn't need to be published, since they might touch off another SST-style brouhaha. NASA remained supportive of their research, however, if scared to death of its potential for public image problems.

12. Harold Johnston held his post despite clandestine political efforts through the board of trustees from within Governor Ronald Reagan's administration to fire the pesky chemist for opposing federal SST policies, and, it was assumed, for being a general environmental nuisance.

13. Rowland remembers it this way: "Yes, we knew that we were going to get a lot of heat, but until you're actually in it, you don't *really* know what it means."

14. The energetic Rowland spent the first week in Vienna renting an apartment, writing the CFC–ozone paper, and simultaneously joining the International Atomic Energy tennis team.

15. The paper was received in mid-January and published about five and half months later. Referee approval came after four months. The rest of the time was prepublication processing.

16. Dr. Paul Crutzen had mentioned the upcoming paper in a talk in Sweden, not realizing that a newspaper reporter was in the audience. On the basis of that reference and a preprinted copy of the Rowland-Molina paper, the reporter wrote an article that headlined the finding. The article got the immediate attention of a Du Pont public relations executive in Europe, who was aghast that Rowland and Molina had used Du Pont's trademark name, "Freon," in connection with what he considered a "ludicrous" theory. After an upset phone call from the man, a chastened Sherry Rowland changed the wording to substitute the generic word "chlorofluorocarbons" for "Freon." In her excellent account of this sequence in the book *The Ozone Crisis,* Sharon Roan reports that in that first phone call, and in subsequent contact, the Du Pont man never seemed to be in the least concerned that the point of the Rowland-Molina paper was, in effect, that his company was producing a product that might be disastrous for mankind.

17. The others were Pennwalt, Union Carbide, Allied Chemical, Racon, and Kaiser Industries. There were a total of twenty producers in the world, but the U.S. group made up about half of world production by volume.

18. In fact, in the sixties, a rash of teenage deaths from "sniffing" or breathing the CFC's in spray cans profoundly upset Du Pont and its senior executives. Though the incidents resulted from purposeful misuse of the product and did not indicate any toxicity, the fact that even the safest of products produced by the most careful of companies could still be tarred with the public image of causing unnecessary death was an excruciating problem.

19. The stratosphere was considered only with respect to global urban smog issues.

20. On the basis of a friend's recommendation, Sherry Rowland sent preprints of the paper from Vienna to Ralph Cicerone and to Paul Crutzen, neither of whom he knew. The key to the calculations was the CFC cross sections in the UV, and this Rowland gave to Cicerone.

21. Scientific journals such as *Nature* and *Science* strongly discourage prepublication publicity by threatening to cancel publication of an article if newspaper articles appear in advance.

22. They also prepared a 150-page support paper.

23. Rowland and Molina also put the expected ozone depletion estimate (with indefinite CFC production continuation at the then-current production levels) at 7 percent to 13 percent. This pair of figures is important because the future calculations were usually done with the same premise: indefinite continuation of CFC emissions at present levels.

24. The media—and especially science writers—had been bombarded with the "end-of-the-world-crisis-of-the-week" for several years, and this seemed like yet another routine doomsday theory.

25. The Wofsy-McElroy findings were not published in *Science* for several more months.

26. The U.S. aerosol market in 1974 had risen to a production level of almost three billion cans annually, about 80 percent with CFC's as the propellant. That works out to around fifteen cans per person per year—sixty per family. When Sherry Rowland and his wife threw their spray cans out in 1973, they found fifteen in their house, and didn't consider themselves big aerosol fans.

27. This was a fact that gave Richard Nixon's friend Robert Abplanalp an additional headache of massive proportions. Abplanalp had just suffered through the demise of the Nixon presidency, and now the basis of his multimillion-dollar fortune was not only being threatened financially, but excoriated as the worst example of environmental rape: the inexpensive plastic valve on the top of each CFC-loaded spray can was manufactured by, or under license from, Abplanalp's Precision Valve Corporation. Abplanalp was so spooked, he took the extraordinary step of writing to the chancellor of the University of California at Irvine asking that Professor Rowland be prevented from further public discussion of the issue. The request was firmly denied.

28. IMOS membership was entirely government employees, one each from about fourteen federal agencies. One of the co-chairmen, Warren Muir, is quoted by Sherry Rowland as saying that he had initially looked on the committee as an opportunity to finally rule against a crackpot kind of environmental alarm. He quickly converted when confronted with the facts, however, and became a strong leader.

29. Chlorine nitrate is very hard to handle because it reacts very rapidly with any water in a lab experimentation system. It took researcher John Spencer several months to learn how to handle it cleanly. When the subject of chlorine nitrate came up later in the midst of Rowland and Molina's ozone-chlorine research, they had a four-month head start on everyone else in being able to test their ideas about the compound as a stable sink for chlorine.

30. The formal rule making was published in the *Federal Register* by the Environmental Protection Agency, the Consumer Product Safety Commission, the Department of Health, Education, and Welfare, and the Food and Drug Administration

Notes [303]

on Friday, March 17, 1978 (Part II). This was the final rule. There was a previous rule requiring warning statements on consumer products that contained CFC's which was published on April 29, 1977 (42 FR 22017) that had gone into effect on October 31, 1977. Yet another Notice of Proposed Rulemaking came out on August 20, 1978, for sunscreen drug products sold over the counter, standardizing the sunscreen ratings and product quality with a numerical rating that we still see on such products today.

31. The book even came after the public—and Ms. Burford—had knowledge of the ozone hole over Antarctica.

CHAPTER 3

1. The Dobson instrument was devised by Professor Dobson in the 1920s. By 1985, most of the Dobson instruments in use around the world were twenty-five to fifty years old, making both obsolescence and worn-out conditions very real problems.

2. There are so-called reservoirs that will lock up chlorine atoms in stable molecules that simply don't react in any appreciable way with ozone molecules (O_3). Two of these are HCL (hydrochloric acid) and $ClONO_2$. Joe Farman had proposed the idea that the reaction $ClO + NO \rightarrow CL + NO_2$ would change the atmospheric balance between NO and NO_2, which would leave greater volumes of free chlorine available to attack ozone. This process would be related to the extremely cold temperatures ($-80°$ C). Susan Solomon realized that such a scenario would work primarily above twenty kilometers, and more precisely at thirty kilometers, which is substantially above the level of most of the ozone layer at that time of year over Antarctica. If the lower atmosphere was involved, then, something else must be happening to cause such a precipitous loss of ozone over such a short period of time.

3. The conference in Switzerland was organized by Bob Watson to put the finishing touches on the 1986 report on "Ozone 1985."

4. Since the consequences of ozone layer depletion can be predicted in human terms by references to increased skin cancer or decreased food production through crop damage—all resulting from increased amounts of UV-B radiation leaking through the depleted ozone shield—the health of the ozone layer as an *issue* appears to hinge on that one measure: how much UV-B gets through to Earth's surface. In fact, that's not the whole story. The Stratosphere can be roughly divided into two zones, upper and lower. The upper Stratosphere (usually above thirty kilometers), and the lower Stratosphere (fifteen to twenty-five kilometers). The effects of CFC–borne chlorine on ozone molecules is not the same in these two upper and lower zones, and this is a key point to understanding the effectiveness of the chemical industry's attacks (among several) in seeming to water down the Rowland-Molina theory.

The mechanism of ozone destruction postulated by Rowland and Molina (that CFC's break down to release free chlorine which steals an oxygen atom from an ozone molecule [O_3] forming ClO, then gives its oxygen atom to another free oxygen atom forming another stable O_2 molecule) works best at about forty kilometers above the Earth, in other words, in the *upper* Stratosphere. At that altitude there are plenty of O atoms for the ClO + O step. The Rowland-Molina mechanism is not very effective *below* thirty kilometers because of the lack of free O atoms.

From the point of view of someone at the surface looking up and measuring total ozone, there is no effective differentiation necessary to measure how much UV-B is impacting the Earth. Yet, since there are at least two zones vertically, it stands to reason that a decrease in the upper Stratospheric content of ozone could be completely counterbalanced by a corresponding *increase* of ozone in the lower Stratosphere. In fact, most of the changes between 1974 and 1985 in the estimates of calculated total ozone while modifying and seeming to water-down the Rowland-

Molina warnings of 1974, were changes in the calculated totals of ozone in the *lower* Stratosphere. The estimates for the *upper* Stratosphere were never really in doubt. There is a legitimate question here: If the sum total at the surface is unchanged, why worry if the balances between upper and lower Stratosphere are changed? In other words, why be concerned if smog *increases* ozone in the lower Stratosphere while CFC–borne chlorine decreases ozone in the upper Stratosphere if the total effect at surface level is roughly the same?

Because, among other concerns, the absorption of UV-B in the upper Stratosphere by ozone is the engine of heating for the Stratosphere. If there is less ozone, there is less heating, and that could change the atmospheric equation in ways we have yet to understand or even postulate. By the same token, changes in the lower Stratosphere also alter the heat flux of the atmosphere.

When total ozone loss due to Rowland-Molina considerations was estimated at 18 percent, the upper Stratospheric loss accounted for perhaps 10 percent of that, and the lower Stratosphere subtracted another 8 percent. When later studies dropped that total-loss estimate to a minus 5 percent, the upper Stratosphere *still* contributed a minus 9 percent, but the lower Stratosphere was now assumed to be gaining 4 percent in ozone content through various reactions. In other words, before 1983, most of the fluctuations in total ozone loss, which began to eat away at the perceived credibility of the Rowland-Molina mechanism and more important at the credibility of the warnings against continued infusion of CFC's, were the result of refinements in the prediction of loss or gain in ozone in the *lower* Stratosphere. In Professor Rowland's own words:

"The loss of ozone in the upper Stratosphere from CFC release has never really been doubted—all of the models showed heavy depletion in the forty-kilometer region. Ozone loss at this altitude will shift the altitude of heat input to the Stratosphere and therefore the *temperature structure* of the Stratosphere itself. This potential source of climatic disturbance was *always* there, even if skin cancer were not so serious because of ozone increases in the lower Stratosphere. The opponents of controls [on CFC's] were generally very effective in deflecting any concerns about the upper Stratosphere ozone loss into a concern solely for *total* ozone."

5. While Rowland and Molina had been careful to state their ozone-loss estimates as a projected *range*, critics found that they could appear to show the Rowland-Molina estimates to be exaggerated by starting with the high end of that range and then graphing the refinements as a slow reduction of the high-end estimate. Lovelock joined the fray and used just that method in *The New Scientist* by graphing the initial loss estimate as 18 percent rather than admitting that Rowland and Molina's real range of estimated loss was 7 percent to 13 percent. Even the National Academy of Sciences used the same shady method in 1984 by starting the Rowland-Molina estimates at a flat 15 percent.

6. The bedrock assumption of science is that there is order in the universe, and that the scientific method can uncover that order—and the truth—by slow, careful, steady experimentation and analysis. Albert Einstein once reaffirmed his belief in that concept by saying that "God does not play dice with the world," yet theoretical and experimental physicists keep unlocking the "last" door to something only to find another door inside. As one physicist studying high-energy particle physics put it, "We get down to what we think is the final, smallest piece of matter, but when we pick it up and shake it, it rattles!"

7. Nimbus-7 carries both the TOMS (Total Ozone Mapping Spectrometer) and SBUV (Solar Backscatter Ultra Violet Radiometer). The TOMS data program was run by Dr. Arlin Krueger, and the SBUV by Don Heath. The TOMS unit gives a finer grid (around 300,000 vertical profiles per day) than the SBUV (around 3,000 per day), but they do not overlap in terms of giving a general picture of ozone concentrations. There is no audible alarm programmed into the computers for strange data parameters even if there had been no programmatic suppression of extremely low

ozone readings. Significant and strange data is simply flagged. Suppressed data isn't even available except on the raw tapes of down-linked satellite transmissions.

8. Don Heath maintains that his team had spotted the extremely low readings several months before Joe Farman's paper, but that they were so wild in comparison with normal ozone levels, that the team wasn't convinced they weren't looking at a degraded satellite instead. Heath claims they were slowly looking at the anomalous readings within their own offices when Farman's information broke. There are others who disagree vociferously with that claim, and are anything but charitable in their excoriation of the NASA team for suppressing any range of Dobson readings to begin with. In any event, the incident was of major importance, and a terrible embarrassment for the Goddard team.

9. Ralph Cicerone did show a nonlinear effect around 1983 in which ClO became increasingly efficient in depleting ozone at some future time when ClO exceeded NO_2, when there was more Cl than could be tied up in the molecule $ClONO_2$. This is actually the Antarctic situation which sets up the nonlinear destruction which causes the ozone hole when NO_2 is removed from the reactive field by being locked up in the ice crystals of the Polar Stratospheric Clouds.

10. This is a rough paraphrase of a brilliantly succinct description of the changeable nature *of* nature—a phrase that comes from the writings of Will Durant: "Civilization," he said, "exists by geologic consent, subject to change without notice." At this early stage in man's attempt to master his environment without destroying it, we would be well advised to think long and hard on the actions civilization has taken that are apparently based on the misguided assumption that Durant was wrong.

11. This "standard" technique included all the data with no changes, and thus mixed both the good and reliable stations with the not-so-good and the total unreliable stations. In addition, such efforts sought a *global* average, which put a heavy load on a few stations in the sparsely covered regions of the world. Since most of the stations on Earth were above latitude 30 degrees N, the readings from the few stations to the south and in the Southern Hemisphere were given enormous weight in arriving at the overall average.

12. This work became Harris's Ph.D. thesis. He received his degree in October 1989.

13. The histories of most breakthroughs in the understanding of a scientific mystery usually have their roots in many elements of research over a long period of time—various "pieces" that at last begin to fit into a defined "puzzle." This was exactly the case with respect to the sudden appearance of the ozone hole in Antarctica, and the major scientific race to figure out what was happening. While heterogeneous reactions on ice crystals in the strange clouds of the Antarctic turned out to be the driving factor, the basic research that led Susan Solomon to the doorstep of the answer in 1974 had been under way for some time, and Sherry Rowland along with Mario Molina and others were deeply involved. While the story was covered very well by Paul Brodeur in "Annals of Chemistry: In the Face of Doubt" (*The New Yorker*, June 9, 1986), and by Sherry Rowland himself in a paper entitled "Chlorofluorocarbons and the Depletion of Stratospheric Ozone" (*American Scientist*, January-February 1989), Professor Rowland's own anecdotal remembrance bears reproduction here:

"In late 1983 I asked myself whether HCl [an atmospheric chlorine sink] would *really* last 3 months without reaction . . . [and] I concluded that this was unlikely, and began thinking about possible reactants for HCl. This quickly brought me back to HCl + Chlorine Nitrate, and then to water plus Chlorine Nitrate once the entry lists had been reopened. I had first talked about HCl + $ClONO_2$ at . . . a weeklong assembly of the original National Academy committee. Harold Schiff and Fred Kaufman had asked for examples of possible reactions which might make the ozone depletion worse, and I gave them HCl + $ClONO_2$ because it took two chlorine reservoirs and reacted them together to re-activate chlorine chains from both. John Spencer even did an experiment and found with our 1975–76 equipment that the reaction was complete by the time he could walk down the hall—three minutes or so. We couldn't do it faster with that equipment, and then in the 1978-79 period with 16-18% calculated ozone depletions, the search for reactions which would make it still worse slackened. Now, however, around January 1984, we returned to HCl and $ClONO_2$ with a Japanese postdoctoral, Haruo Sato, and different equipment. He quickly (May) showed that both HCl and $HlONO_2$ were gone in one or two seconds! Further, the reaction of water with $ClONO_2$ was also fast. In June, 1984, a stratospheric ozone meeting was organized by Bob Watson at Feldafing, a West German village near Munich. In preparation for this meeting, I calculated approximate gas kinetic rate constants for these two reactions—that is, we said, "We don't know whether these reactions are occurring in the gas phase or on surfaces, but we'll calculate how fast it has to be per collision on the assumption that it is all gas phase. Then, I asked Don Wuebbles to put these rate constants into his model calculation to see whether they made any difference. In effect, if these reactions were actually gas phase at these rapid rates, *and* hardly affected the ozone depletion calculations, then it would probably not be worthwhile to try to find out whether they were gas phase reactions or surface processes. The Wuebbles calculations are shown in Figure 6, *American Scientist* [article cited above]. These rates made *enormous* changes in the calculations. Instead of an eventual ozone depletion of 4.2%, the HCl + $ClONO_2$ at our rate constant gave a depletion of 31.7%!

"I then made a presentation at the Feldafing meeting of these results, essentially alerting the community to the potential influence of these reactions on ozone depletion calculations if they were actually occurring in the Stratosphere. About this time (1984) Paul Brodeur interviewed me again for his New Yorker article which came out in June, 1986 (he couldn't get it out in 1982 or in 1984, but the ozone hole sprung it past the editor in 1986). I told Paul about these reactions of chlorine nitrate, and speculated that surface reactions might be occurring on particulate volcanic debris.

"Sato and I did not publish our results in 1984, other than the Feldafing presentation, but [we] did publish in 1986.

"During the second half of 1984, a number of research groups took up these reactions with two general problems:

"1. What was the real gas-phase rate?

"2. What was the efficiency of the reactions on various surfaces?

"During late 1984, early 1985, when I mentioned the possibility of surface reactions, the answer often came back (usually from industry) that the efficiency on surfaces was only 1 in one million, and that couldn't be fast enough to be important in the real atmosphere.

"The surface problem is a standard one in gas phase kinetics, and the lore provides a small catalog of usually 'inert' surfaces. None of these made the HCl + $ClONO_2$ reaction appreciably slower for us, as we reported in 1986. Scientists at the University of California Riverside with very large vessels (smog chambers) which have very low surface/volume ratios compared to ours showed that the real gas reaction was very, very slow and was certainly not important—which left surface reactions.

"Meanwhile, Susan [Solomon] and Rolando Garcia were developing a two-dimensional model, as was Don Wuebbles at Livermore. In the 1-D models, latitude is not a variable, so there is no possibility of predicting an effect localized in the south polar region. With a 2-D model, polar and temperate predictions can be different. In Susan's 2-D model, she tried for a polar ozone loss with the standard reactions and clearly could *not* get an ozone hole. However, when she put in either HCl or H_2O + $ClONO_2$ at our hypothetical gaseous rates, the hole appeared. This was in effect a surface-catalyzed simulation *if* the surface efficiency was not 1 in 1,000,000, but perhaps 1 in 100. She and Garcia distributed a two-author preprint, which brought Wuebbles into the picture because he had a 2-D model and was putting in 'our' reactions, too. In the end, all four of us participated, but the driving force in the calculation was Susan Solomon."

Notes [307]

14. The paper by these four was entitled "On the Depletion of Antarctic Ozone" and was printed in *Nature* 321, no. 6072 (19 June 1986): 755–58.
15. Basically, the PSC's do two things:
    1. Remove NO_2.
    2. Change HCl and $ClONO_2$ into Cl_2 which splits in the rays of the first sunlight of the Austral dawn. In effect, it converts chlorine reservoirs—sinks—into chain-active chlorine ready to gobble ozone.
16. Molina had published a paper on such reactions before the Antarctic ozone hole "appeared" in Joe Farman's report.
17. It had been assumed for years that hundreds of thousands of collisions between molecules would be necessary before the reactions would occur on the ice surfaces, but in fact only a few were needed.
18. These findings were reported by M. J. Molina, T. L. Tso, L. T. Molina, and F. C. Wang in "The Chemical Interaction of $ClONO_2$, HCl and Ice, and the Release of Active Chlorine in the Antarctic Stratosphere," *Science* 238 (27 November 1987):1253–59.

In this same issue of *Science,* a second paper by investigators from SRI International in Menlo Park reported similar findings which supported Molina's conclusions.
19. OClO is the pilot fish that betrays the presence of the ozone-eating shark, ClO. ClO could also be described metaphorically as the smoking gun, what's left after an ozone molecule has been eliminated. Chlorine *di*oxide, OClO, is a by-product of some nonozone-destroying reactions. But showing the presence of *both* OClO and ClO in the same reactive conditions goes a long way toward telling the entire tale. Thus the need for an instrument to "do OClO."
20. The microwave technique for detecting ClO takes advantage of "pressure broadening" of the ClO signal which is different at different altitudes. Thus the readings can be broken down to show the concentration of the molecules at various altitudes, yet the readings are all taken from the same location.

CHAPTER 4

1. There was more than one theory centering on dynamics as the root explanation. One concerned the transport of ozone from and to the Troposphere; another considered transport of ozone from the mid-latitude Stratosphere.
2. In some respects it was not at all surprising that the four select teams on the ice for NOZE-1 had been successful the first time out. Ozone-related measurements had become a highly competitive business as far back as 1971, and under the heat of industry (and sometimes governmental) pressure, the field had matured during a fifteen-year trial-by-fire. During that time, no ozone measurements were beyond intense professional scrutiny, and thus the survivors of that weeding out process had to be very, very good. Certainly good fortune played a substantial role during the first days of NOZE-1, but the teams involved were made up of only world-class experimental scientists.
3. Months later the same reporter called Susan at her office in Boulder, and the first words out of his mouth were: "Do you have any idea how hard it was to place that call? And you wouldn't even accept it!" There was no acknowledgment of the trauma he put her through for two hours, or of the fact that he had grossly misrepresented himself to get through a restricted line.
4. Susan Solomon disagrees with this premise, feeling that "there would have been 'fire' no matter how it was released." Indeed, those who favored a more dynamical solution as an initial postulate explanation could not have been expected to roll over in silence in the face of an expedition's findings, however quiet the release. To fail to challenge such findings energetically—and perhaps even publicly—would seem

a tacit endorsement. Nevertheless it seems to the author that one public airing of such an issue *absolutely requires* a balancing rejoinder from researchers with a different point of view, and this instance seems to have followed that logic.

5. Actually, very few of the NOZE-1 team were actually chemists. Barney Farmer is a physicist, Bob de Zafra is a physicist, Phil Solomon (no relation to Susan) is an astronomer, and not even Dave Hofmann is a chemist. Susan Solomon, of course, is, and Susan was the leader.

CHAPTER 5

1. Peter Wilkniss was also the scientist who led a 1972–73 research effort which was the second to measure CFC concentrations in the Troposphere. James Lovelock was the first.
2. This was the report under discussion at Les Diablerets, Switzerland, in July 1985.
3. In 1988 he finally, formally became a NASA employee. Previously he was "on loan" from JPL.
4. Reports came from NASA, World Meteorological Organization, United Nations Environmental Programme, the British Government, National Academy of Sciences/National Research Council, and the Environmental Commission of the European Community.
5. This report, however, also quoted the statistical studies that said Stratospheric ozone levels worldwide had not changed as measured by the Dobson Network.
6. The United States and Canada and the Scandinavians had pushed very hard for a protocol to be signed at the same time in Vienna banning aerosols worldwide and other specific uses. After months of bitter wrangling, the Toronto Group (as they had been labeled) lost out, and only the framework convention itself was finalized. But in the following years, according to Bob Watson, it became very clear that to have signed such a protocol at that time could have been disastrous in that it would have given the international community the false sense of security that enough had already been done even after the ozone hole had been detected and its source identified.
7. According to Sharon Roan in her book *Ozone Crisis* (New York: Wiley, 1989), Du Pont chemist and spokesman Donald Strobach admitted to environmentalists in 1986 that they had discontinued the search in 1980, "so confident were they that the CFC-ozone depletion theory had died for good. The company was, in fact, building new production facilities in Japan and was planning a move into China." Sherry Rowland, however, knew that the chemical companies were well aware of the potential source of viable substitutes, the so-called HCFC's, which will break down in the Troposphere releasing their chlorine atoms at safer altitudes. The distinction between CFC's and HCFC's in the atmosphere was pointed out by Rowland in 1974! At the same time, Professor Rowland specifically made the point that substitutes for CFC's could be found among the HCFC class.
8. There were 280 parts per million in 1850, and today the atmospheric concentration stands at 50 parts per million.
9. As a government witness, Bob Watson had to have his statements written and approved by superiors in advance. Sherry Rowland was not hamstrung by the same limitation.
10. Hansen's interests had been the atmospheres of neighboring planets, and from within his research position he had quietly worked throughout the seventies on modeling the atmosphere of Venus, helping to plan the spacecraft programs that had—and would—fly to the neighboring planet to sample its atmosphere, probe its surface and atmosphere, and radio its data back to an eager team of specialists for whom every clue would be a precious jewel.

Somewhere in the late seventies, however, Jim Hansen had looked homeward, becoming aware that the abstract calculations and postulations had some real-time application to his own planet's atmosphere. The same greenhouse effect that keeps Venus cooking at the surface is a warm blanket to the Earth, and Hansen became interested in applying the same principles to models of our atmosphere. There were three key scientific questions: How large is the man-made Greenhouse Effect going to be, when will it exceed the magnitude of natural climate variability, and what effects will it have on man?

The results startled him, and led him and his colleagues at GISS to publish a paper in *Science* in 1981 that contained a dramatic conclusion that Earth's atmosphere was in fact warming up, and the amount of warming was consistent with the measured increase in greenhouse gases. The conclusion was surprising, and it instantly sparked scientific controversy, since earlier data showed that the Northern Hemisphere had actually cooled off between 1940 and 1970. Many scientists had used that less-than-complete data as a basis to argue that the Earth was now headed into an ice age. It was Hansen's 1981 paper that led to his appearance before Representative Gore's committee in early 1982, a hearing Gore opened with the declaration that the Greenhouse Effect was no longer a matter of concern only to the "sandals and granola" crowd.

11. The simulation was the GISS 3-D climate model.

12. The testimony OMB deleted had read:

"Finally, I would like to point out that the supply of the two key ingredients must be increased in order that we can use the observational data to quantify the greenhouse effect and develop adequate climate models. One key ingredient needed is an influx of young scientists with appropriate training; we must begin training students now if we are to have scientists available in the next decade when the need for information increases. The second key ingredient is research funding. Progress in developing more realistic climate models is seriously affected by continued decreases in funding for that research. More and more time is spent pursuing smaller and smaller grants, with a large negative impact on productivity."

Hansen was told he could make an oral statement to this effect, but only if he carefully qualified it as his personal opinion. If he did, however, it would take up a portion of the five minutes allotted for the rest of his testimony.

13. Only one scenario for trace gas growth had been completed as yet, and with the capability of the 1975-vintage Amdahl computer at GISS (described by *The Philadelphia Inquirer* as the equivalent of a 1950 DeSoto), it would be at least another year or two before a broader set of scenarios could be examined.

14. The movie showed every day in October for each progressive year starting in 1979 and running through 1985 as projected on the south polar region by computer simulation using actual TOMS data. The work had been done by Mark Schoeberl acting on the request of Richard Stolarski, Don Heath (who ran the SBUV, Backscattered Ultraviolet instrument), and Arlin Krueger (who created and ran the TOMS instrument). They had used the sequence in many a conference since the discovery of the suppressed TOMS data regarding the existence of the ozone hole.

15. The other scientific sampling packages that were to fly aboard the ER-2 had already been installed earlier and used for other missions, including a 1987 mission to Australia (the Stratosphere/Tropopause exchange study). Anderson's, however, was a completely new addition and engineering challenge both to build and to install aboard the ER-2.

16. Technically, the retrofitted DC-8 became a Model-73 with the new engines and new air-conditioning/pressurization system.

17. The Alliance came into being only about 1980, replacing other groups that were formed in the 1974–75 time frame when the chemical industry perceived itself under environmentalist attack.

[310] *Notes*

CHAPTER 6

1. There are a scattering of islands along this route, but none would be very useful for a forced landing.

2. Because of the importance of ClO detection to the expedition, it is tempting to reach the conclusion that Jim Anderson's instrument (for ClO/BrO detection) was the only one of significance working aboard the ER-2, but that is wholly misleading, as numerous colleagues of Anderson and Anderson himself have hastened to point out. There was, in fact, an important array of various detection experiments aboard the aircraft, including instruments to detect the levels of O_3, NO_y, H_2O, N_2O, and particle instruments, all of which were of extreme importance to developing a scientifically complete picture of what was causing the Antarctic ozone losses.

3. In mid-August, the edge of the vortex extends north to latitude 65 degrees S. The terminator—the dividing line between perpetual night and diurnal sunlight—is at latitude 76 degrees S.

4. The fact that the air was denitrified was discovered by David Fahey after the first three ER-2 flights had been completed and the data reduced.

5. The data reduction and transfer to Punta Arenas required a total elapsed time of no more than four hours between on-board satellite observation and receipt of the final product at Punta Arenas.

6. There had already been some ozone loss at the edges of the vortex as early as mid-August, but that fact was not known at the start of the expedition.

7. Just how effective these winds are in "walling off" the interior of the vortex is still a highly controversial issue within the atmospheric science community. Even the use of terms such as "walling off" and "containment vessel" will spark immediate debate among scientists who are otherwise in fairly close agreement on the nature of the ozone loss mechanism over Antarctica. There is, in fact, a fairly widespread consensus that the winds (which reach 150 knots at sixty-eight thousand feet around the perimeter of the vortex) do tend to restrain the interior air mass from free mixing with the air of lower latitudes. The extent to which this tendency becomes a segregation of the air within the vortex, rather than merely an impedance to completely free mixing, is the nexus of the question.

8. The descriptive pilot notes of the cloud formations are fascinating. Those written during the second flight on August 18, 1987, with Jim Barrilleaux at the controls, are as follows:

"1. First noted white lenticular clouds on the horizon when approaching 60° S. Passed abeam of northern edge of lenticular clouds at 62° 30' (approximate latitude of Morsh Station). Clouds were east of flight track, estimated to be over the peninsula. Clouds appeared to vary in intensity from faint to optically dense white clouds. As I flew to the southernmost point, the sky coverage increased from approximately 20% initially to 50% coverage, also the percentage of optically dense clouds increased from about ⅓ of total at the north end to ½ of total. Clouds were dense at the southern end. Clouds appeared to be at least 5,000' to as much as 15,000' above me.

"2. Observed a large, dense, light grey colored cloud beneath the roll clouds. This large cloud extended as far east and south as I could see. The upper surface of this cloud was extremely smooth—like a "roll" or lenticular cloud. The cloud was below the ER-2, estimated somewhere in the mid 50,000' range (guess).

"3. Approaching the peninsula, also noted a thin haze layer when looking to the west (right wing) and south (nose). Unable to observe to the west (left wing) due to brightness of clouds over peninsula. Haze layer seemed to extend +/− 5,000' (crude guess) from ER-2 altitude. As aircraft flew south, haze became slightly thicker. Also observed slight variations or layers in haze at southern end of flight track (65–67° South). Haze not as noticeable on north bound flight as soon was right above horizon and frequently shined into my eyes."

9. Nevertheless, the ClO concentrations were much more abundant than midlatitude summertime measurements at equivalent altitudes. ClO is either being produced by PSC's outside of the vortex edge, or is being "spun off" by vortex erosion (or both).

CHAPTER 7

1. ClO profiles were reported from both NOZE-1 and NOZE-2 by Bob de Zafra and Phil Coleman, but the instrumentation on NOZE-2 enabled more precise measurements.
2. The events of September 5 would eventually lead to the conclusion that the miniholes were the immediate result of dynamics—air movement—though the propensity for ozone loss was a chemical phenomenon associated with PSC's.
3. This is now the key question: the "containment vessel" versus the "flow reactor" issue.

CHAPTER 8

1. Adrian Tuck, who shaved his mustache sometime later, characterized it as more of a "Mexican bandit" style rather than the author's firsthand description of "walrus." He does admit, however, to being barrel-chested.
2. The degradation of the diffuser plate was from exposure to the sun, and it was a well-established fact as early as 1980. The ozone calculations, in other words, were adjusted to compensate for the downward drift of the readings. The question before the Trends Panel members, however, was whether the algorithm which Arlin Krueger and Don Heath were using to adjust for the drift was nevertheless still leaving a bias in the adjusted data.
3. The evolution of the ideas that later coalesced into a theory to explain the Antarctic ozone destruction by means of heterogeneous reactions on the surfaces of ice crystals began long before Joe Farman's paper was published. (Please see note 13 in Chapter 3 for a fascinating look at the history of this theory and its evolution.)
4. The extremely detailed calculations they were involved in producing were not just a reevaluation of previous work, but an entirely new statistical calculation which was threatening to overturn a long-held conclusion arrived at years before by professional statisticians: that there was no ozone loss trend found in the Dobson station readings. Du Pont's McFarland, for one, wanted to be absolutely certain that the potential conclusion that there *was* in fact a statistically significant loss of ozone in the Northern Hemisphere didn't just originate with the 1982–83 results which had been presumably affected by El Chichón.
5. The resulting paper, "Observations of the Nighttime Abundance of OClO in the Winter Stratosphere above Thule, Greenland," by Susan Solomon, G. H. Mount, R. W. Sanders, R. O. Jakoubek, and A. L. Schmeltekopf, was published in *Science* on October 28, 1988. With the data taken in February 1988, the paper written by and submitted in July with an October publication date, that was about as fast as a scientific team could report through the refereed journals.
   In the 1989 expedition, Jim Anderson would find as much OClO over the Arctic as over the Antarctic.
6. While the worldwide Dobson network measures a ratio of UV-A (which is not absorbed by ozone) to UV-B, the direct-reading UV-B meters measure only UV-B light striking the sensor from any angle. Thus, while decreased Stratospheric ozone would be detected by a Dobson even in the presence of increased ozone presence

[312] Notes

around the reading site, a UV-B meter would simply average the Stratospheric loss of ozone (as increased UV-B) with a low-altitude gain in ozone (due to smog and pollution which would absorb the added UV-B reaching the surface), and perhaps report a steady level of UV-B which would falsely imply that there had been no change in Stratospheric ozone.

7. At one point, Graham was barely prevented from pouring twenty million dollars into an off-the-wall scheme to destroy the CFC's in the atmosphere by stimulating the ionosphere with radar! Even if enough radar energy could be focused on the ionosphere to energize such a process, there is a rather substantial flaw in the plan: CFC's never reach the ionosphere!

8. The previous statistical approach had assumed that if an ozone loss were to occur, it would be the same for all seasons of the year. When that assumption is coupled with the realization that winter Dobson readings are extremely "noisy" compared to more stable summer readings, then it makes sense to weight the data so that the more stable summer readings are predominant. If it's the same for all seasons, this is a sound methodology. If there is any difference in terms of seasonal ozone loss, this method can yield significant errors.

The actual summer readings do, in fact, show at the most very small changes in ozone levels, and so this had become the statistical conclusion for the entire data set from the world's Dobson stations. *But,* the winter figures were showing significant losses. Yet those losses were hidden by the use of the stable summer data as the predominant authority.

Another aspect of the problem was the extensive delay in publishing what data existed. The 1986 WMO-UNER report, for instance, was quoting published evaluations of ozone data through 1979–80 because updates incorporating 1984 and 1985 figures had not been worked up for publication.

9. This is a controversial issue. Another view is that the Antarctic vortex is more of a flow reactor than a containment vessel, and as a flow reactor, it spins off ozone-poor air to the mid-latitudes. In other words, mid-latitude ozone losses might be more the result of mixing ozone-poor polar air with mid-latitude air than the result of any significant increase in mid-latitude heterogeneous reactions.

10. Chairman Heckert sent Senators Baucus and Durenberger a second letter justifying his previous letter by saying that the March 15 Trends Panel report was new evidence that simply updated the previous evidence, ignoring the fact that the previous "scientific evidence" was not considered valid. The letter, as quoted by Sharon Roan in her book *Ozone Crisis,* ended with: "While we believe the short-term risks to health and the environment from CFC's is negligible, we nonetheless have concluded that additional actions should be taken for long-term protection of the ozone layer." In addition, Heckert, for inexplicable reasons, excluded Senator Stafford from the letter's distribution list, and from any personal visits by Du Pont representatives. The small-minded snub was shrugged off by Stafford, but well-noted by his colleagues.

11. There is another aspect of "missed-opportunity" to this reticent attitude as well: Any other CFC company could have had a ten-year jump on Du Pont in the HCFC replacement fluorocarbon market if they had either been more environmentally conscious, or more astute as marketers.

12. Compounding the time problems was the fact that while some scientists have clear voices and can think and speak with sparkling clarity on their feet, public speaking is not a required subject in a Ph.D. curriculum, and some of the presenters simply couldn't communicate rapidly or clearly, leading to additional questions and lost time.

13. According to Adrian Tuck, the "cornfield meet" was really over the question of whether or not the Antarctic vortex is better approximated as a flow reactor (the contention chiefly of Tuck, Mike Proffitt, Dan Murphy, and Ed Danielsen), or as a containment vessel (the contention of Dennis Hartmann, Mark Schoeberl, and as

assumed by Jim Anderson for his kinetics analysis). The confrontation is still ongoing—there was a major clash over it at the AASE meeting at Charlottesville, Virginia, in June 1990. The next Antarctic mission, in fact, will probably be targeted to address this question.

14. Jagged, irregular surfaces are even more effective than flat surfaces—such as ice flakes rather than large crystals.

15. Sherry Rowland and Neil Harris would later do a serious study of this subject in conjunction with the readings from Arosa, Switzerland, which would make a case for heterogeneous reactive ozone losses caused by volcanic debris acting as a catalyst.

16. Maggie Tolbert was not the first scientist to tackle what is still an unresolved issue. According to Mario Molina (as he told his audience at Snowmass), it is clear that sulfuric acid particles are rather *inactive* at lower latitudes. The question is whether they become activated *outside* the polar stratosphere before they are transformed in the polar regions to PSC's—Polar Stratospheric Clouds.

17. To be precise, the Rowland and Molina predictions concerned potentially catastrophic ozone losses in the upper stratosphere at lower latitudes in the middle of the next century. What Snowmass validated through the findings of NOZE-1 and -2 and the Punta Arenas expedition was catastrophic ozone loss in the lower Stratosphere at high latitudes in the last half of *this* century caused by a never-postulated permutation of chlorine loading of the atmosphere. Yet the most valuable element of the Rowland and Molina postulations of the mid-seventies—from a societal point of view—is that an alarm was raised at that early date for the first time over CFCs and their potentially destructive effect on Earth's ozone layer. Snowmass did validate that connection—that CFC chlorine loading has now been proven to have had a significant effect on the ozone layer by whatever means at whatever altitude and latitude. CFC's have placed the chlorine in the Antarctic stratosphere in such concentrations that the ozone hole was made possible, and for CFC producers, there was and is no way around that reality. Before NOZE-1 and -2, and Punta Arenas followed by Snowmass, there was still some defiance left in the CFC producers as they continued to blindly maintain that there was no proof of a smoking gun connection between their product and negative planetary impact. Snowmass put the smoking gun on display.

18. In fact, the 1989 Arctic Ozone Expedition showed that ozone depletion was occurring at a rate of 1 percent *per day,* but the meteorology broke it up before heavy losses could occur. With even more chlorine headed toward the stratosphere, the rate of ozone loss by the year 2000 is expected to be considerably faster. Air, with substantially depleted ozone, drifting over the populated regions of the north is an even more probable legacy of CFC production.

CHAPTER 9

1. His Ph.D. research on the atmosphere of Venus had evolved on the strength of a suggestion by Professor James Van Allen, the head of the physics department at the University of Iowa, and the discoverer of the Earth's radiation belt (called the Van Allen belt). Why not, Van Allen suggested, examine the puzzle of how Venus could be as hot as radio wave data from the planet seemed to indicate. No space probes from Earth had yet been sent to the second planet from the sun, and there was a race of sorts to construct theories about Venus that might later be proven by actual spacecraft sampling.

The most popular explanation for the hot Venusian surface was a super greenhouse effect, but Hansen needed to make and defend an original proposition to qualify for his doctorate, and the greenhouse effect had been taken. In its place, he proposed the far-out idea that the atmosphere of Venus was filled with dust which in turn

[314] *Notes*

trapped the solar heat radiating back from the surface. The theory was inventive, but in 1969 a Soviet spacecraft arrived at Venus and radioed back data that utterly collapsed the dust theory. Venus, it seemed, was indeed a greenhouse, and a runaway one at that. A massive blanket of carbon dioxide ninety times denser than that of Earth's atmosphere had permitted the surface temperature of the planet to rise to hundreds of degrees.

The study had paid off for Jim Hansen, however, with more than his Ph.D. By then he had turned to developing computer programs to calculate the effect of the greenhouse gases as well as aerosols (dust) and their effects on the planet's temperature. The exciting thing was the correlation between the amount of CO_2 the Soviets had measured on Venus and the high temperature. It made life on Earth seem even more interesting, since Earth's thinner blanket of CO_2 was presumed to keep surface temperature livable, and yet the thickness of that blanket was being increased every year.

Through the 1970s Jim Hansen continued to work at GISS (while joining Columbia's faculty in 1978 as an adjunct professor of geological sciences) as they planned the design of an instrument for the *Pioneer* spacecraft to carry to the clouds of Venus. By the time *Pioneer* was finally launched in 1978, however, Hansen's interests were focused almost entirely on Earth studies.

Later it would be discovered that Mars' thin CO_2 envelope raised that planet's surface temperature by only a few degrees.

2. In terms of producing well-educated students, the public education system of the United States is broken. From elementary through middle schools and high school level, the system (with some exceptions of course) is abysmally substandard when compared to the education systems of other industrialized nations. But in no area is that truer than the sciences. The state of American scientific education below the collegiate level is nothing short of a national disaster and a national shame with dire consequences for the nation's future. Similarly, Canada, the United Kingdom, France, Scandinavia, and all of the civilized world are increasingly producing two-legged consumers of science and technology who are increasingly devoid of the basic high school level scientific training required to understand the general nature of the processes and the principles involved. Even the crop of adept young scientists in the university pipelines and advanced-degree programs is shrinking—especially in the United States—largely due to the shortsighted societal failure to fund and support the education of young scientists, the all-too-rare programs such as the one from NASA that nurtured James Hansen through his doctoral and postdoctoral preparation.

The point has not been lost on the competitive minds of Japan and Germany industry, and the results can be read in part in the American trade deficits, much of it representative of the declining American capability and willingness to fund research and development or fund the education of the scientists and engineers we blithely expect to *do* the advanced research necessary to develop the future technology-based or science-based products the world is increasingly hungry to purchase.

A by-product of all this shortsightedness is an increasing inability of the average American to look beyond the headlines if they concern science, and a willingness to accept at face value the oversimplifications and homogenized versions of scientific statements or capsulized summaries of scientific papers that may have taken years to research and produce. "I'm not a scientist. I'm not supposed to understand such things" is a viable excuse for the layman confronted with hard-core scientific facts and conclusions, but a rudimentary understanding—and curiosity—is vital to the society. When people lack the basic knowledge of how to ask the right questions or be sensitive to the subtler elements of the answers, an increasingly technological society is in dire trouble, because the ultimate result of general scientific ignorance is the creation of a scientific and technological elite, a "priest" class of scientists who increasingly hold their remote and mysterious truths sacred and at arm's length from the average citizen who can't understand them. Even a minor societal drift in that *direction* is dangerous in a representative democracy, because a decreasing level of public scientific and technical "literacy" (for want of a better term) means an increasing inability of the body politic to make intelligent, informed, independent decisions when science and technology are involved. How, for instance, can a nation decide democratically how to handle the complex legal problems raised by genetic manipulation, and perhaps human cloning, if even the basics of the science are beyond the comprehension of the population? A sure way to dangerously limit democratic choice is to render citizens so ignorant that they must rely on the scientists and technicians themselves to make their choices for them. Unfortunately, the United States of America is drifting in just that direction.

Of course, few scientists would expect the lay public to speak and understand their language, or expect television newscasters to be able to do a thorough and careful job of explaining the mind-boggling intricacies of arcane scientific findings to the unscientifically trained ears. But while the scientists have a responsibility to be clear and precise in *their* communications with the public, the public has a responsibility to be at least minimally educated in how to listen to the scientists, and what questions to ask.

At a minimum, the average citizen should never lose sight of the fact that a true scientist is seldom 100 percent sure of *anything*, except perhaps where to go for lunch. Thus, news reports that thunder in with the definitive report that a NASA scientist is 99 percent sure the heat wave of 1988 is the result of global warming should be taken with a grain of salt the size of Gibraltar until more is heard—or read. After all, that's quite a claim. Perhaps the scientist failed to make himself clear. Perhaps he misled the press, or the press misled him, or they mutually failed to communicate. In any event, normal caution should have warned most people who heard and read of the Wirth hearing and Hansen's statements that few things in life are really that simple, *especially* where cutting edge scientific theories are concerned. There *had* to be more to it.

3. When Jim Hansen was questioned later about his testimony and the tumultuous reaction to it, he defended his words, saying that even with the benefit of hindsight, he did not feel that he would change anything if he had it to do over again. "I understand the position of those scientists who feel I went too far in my testimony, but I just don't agree with them."

CHAPTER 10

1. This quote and the story appear in Stephen Schneider's book *Global Warming* (San Francisco: Sierra Club Books, 1989) on pp. 200–201, and are used with his permission.

2. Dr. Margaret Mead had become involved in man's impact on the climate by 1975, and had convened a conference to find ways to deal with the buildup of atmospheric CO_2. In 1977 she co-authored a book with Will Kellogg calling for an international "Law of the Air" (W. W. Kellogg and M. Mead, *The Atmosphere: Endangered and Endangering*. No. NIH 77-1065).

3. Quote taken from Schneider, *Global Warming*, p. 204, by permission.

4. If there was no air in the enclosure at all, much less of the heat would be retained. In a normal greenhouse, though, the molecules of air inside soak up the heat radiated from the plants and other objects as *they* heat up.

5. True, the simple act of isolating an air mass within a visibly transparent structure keeps the air from dynamical mixing with the outside air, which may be at much lower temperatures. By simple physical retention of the warmed gas, the green-

house's interior can heat up cumulatively. Depending on size, this may be the dominant effect, but in larger greenhouses the well-proven absorption of infrared energy by water vapor, carbon dioxide, and the other trace gases materially aids the heating process, and the infrared-opaque nature of the glass aids its retention.

6. This quote is from an interview of Dr. Roger Revelle (one of the giants in oceanography, and one of the many accomplished world-class scientists who helped organize and conduct the IGY) by author Jonathan Weiner in his book *The Next One Hundred Years* (New York: Bantam, 1990), p. 31.

7. The book was called *Worlds in the Making,* trans. H. Borns (New York: Harper and Brothers Publishers, 1908).

8. The finding was that only half of the extra CO_2 in the atmosphere would be absorbed by seawater because of a built-in buffer mechanism. The finding was reported in a paper by Roger Revelle and Hans Suess entitled "Carbon Dioxide Exchange Between the Atmosphere and the Ocean and the Question of an Increase in Atmospheric CO_2 During the Past Decades," *Tellus* 9 (1957).

9. From Weiner, *The Next One Hundred Years.* p. 64.

10. The average temperature of the planet does have to be inferred from various sophisticated measurements of isotopic decay, but the majority of the data is direct reading from the air itself and the analysis of its molecular content.

11. And that, of course, was becoming more than obvious from the testimony. The need for substantial funding of the advance modeling programs at GISS, NCAR, Fluid Dynamics, and elsewhere is acute—the "dirty crystal ball" needs polishing, and it's the only window to the future effects of the greenhouse gases above us.

CHAPTER 11

1. *The New York Times,* 3 July 1989.

2. Indeed there is some evidence that the atmospheric science community may be disagreeing in public and agreeing in private. In the June 2, 1989, edition of *Science,* Richard Kerr—in discussing a spring workshop on greenhouse-gas-induced climatic change—quoted Dr. Stanley Grotch of Lawrence Livermore National Lab as saying that "if there were a secret ballot at this meeting on the question, most people would say the greenhouse warming is probably [here]." Yet that very workshop was replete with criticism of Jim Hansen's points that there is a warming trend that has been detected, and that trend is linked to the greenhouse effect—a conclusion some colleagues found unforgivable.

3. Wade was relying in part on a thoroughly discounted American Cancer Society report of 1987 as if there had been universal scientific acceptance of its premise that ultraviolet light levels were decreasing in the United States (without accounting for the urban, airport location of the readings, and the role of photocarbon smog and pollutants during the study period). The galvanizing aspect of the Ozone Trends Panel report of 1988 was that there had *already been* a significant and dangerous loss of ozone in the Earth's ozone layer regardless of what might be happening at the poles (the Antarctic and Arctic ozone hole problem), and that means literally by definition that there is more ultraviolet light shining down on all parts of the North American continent, where one obviously limited study validates it or not. Yet the implication of the editorial is that in the ozone debate no damage had been proven, but it is well-established scientific fact that UV-B radiation—which is now increasing in direct linear inverse proportion to the decrease in the stratospheric ozone shield—causes breakdowns in the DNA chromozonal linkages that are the building blocks of carbon-based life on Earth, and that such damage will—not just can—cause a variety of undesirable and serious biological effects including increases in skin cancer and crop damage, in addition to nonlife form damage to paints and plastics. Nicholas

Wade stands firm in his view that the authors of the American Cancer Society report have adequately answered the criticisms of their finding, but that optimism is completely unaccepted in the atmospheric science community. He also ignored the fact that instead of "preventing further thinning" of the ozone layer (something in any event the Montreal Protocol is helpless to do for at least fifty years since the CFC's already released will not even reach the stratosphere completely before that time), the fact is a 50 percent reduction by the year 2000 is wholly inadequate to even slow the progressive damage. In addition, Wade discounted the tawdry history of delay in the ozone debate which ground on for fourteen years while esoteric uncertainties were used as justification for inaction. Yet, for fourteen years the international community could not agree on even the half-effective Montreal Protocol, until the Antarctic Ozone hole was discovered. While it is true that Richard Benedick and Bob Watson were scrupulously careful not to beat the drum about Antarctica in the Montreal negotiations—fearing a backlash if chlorine was not the cause—there was no conscious member of that international assemblage unaware of the implications. If it takes fourteen years and major, perhaps irreversible damage to the ozone shield to get a protocol that is, at best, only partly effective, it is difficult to understand an essayist's optimism that the nations of the world can "presumably do the same with greenhouse" in any timely fashion before too much atmospheric and environmental damage has been done. This is not to beat up on Nicholas Wade in particular. In many respects, his editorial was a responsible voicing of revulsion at an extreme environmentalist's statement that in order to get congressional action, "We'll just have to wait for a hot summer." But in writing the piece, his reliance on discounted "findings" on UV-B light, his overconfidence that the global community can find ways to act in time, and his unintended linkage of Jim Hansen with the concept that the summer of 1988 was caused by the greenhouse effect, were all unfortunate. In fact, for anyone really crying wolf, or overstating the case of greenhouse/global warming to achieve political action through shock treatment, intellectual dishonesty even for a good cause is still intellectual dishonesty, and Wade's point that such stupidity is not needed (and may be counterproductive) is good advice. But few people read editorials with sufficient care to decipher all the nuances of the writer's carefully crafted statement, and therefore the author must write for the audience, not just for intellectual and literary purity. If an otherwise perfectly correct series of statements has the unfortunate effect of tarring someone with the wrong brush, it needs revision. In this case, Jim Hansen got a can of paint in the face. Unfortunately, this is very representative of how the media, and the public, can miscommunicate and misunderstand what the scientists are saying when they try to break a scientific debate into a two-sided issue—in this case the environmentalists against the mainstream scientific community. The delineations simply are not that clear-cut, nor are the statements of the environmental community in any way uniform. To lump organizations as wildly diverse as the respected Environmental Defense Fund on one hand in with a lunatic fringe organization such as Earth First by the unicameral use of the word "environmentalist" is as dangerous and misleading as lumping all scientific opinion in a scientific debate into a homogenized statement characterized by the phrase: "Scientists say that ..."

4. K. E. Trenberth, G. W. Branstator, and P. A. Arkin, "Origins of the 1988 North American Drought," *Science* 242 (1988):167–75.

5. There is a comedic slogan that expresses this oversimplification very well, and it has been seen tacked to bulletin boards and in one instance neatly framed on the wall of a government psychologist in Washington, D.C.: FOR EVERY Ph.D., THERE IS AN EQUAL AND OPPOSITE Ph.D.

6. The GISS researchers had checked available diagnostic data related to droughts, storms, and floods from the climate models of other laboratories and verified that the results were qualitatively similar. Hansen planned to emphasize in whatever he told the congressional committee that his conclusions did *not* depend

upon the ability or inability of the climate models to simulate climate changes in specific regions of the world. This was an important point, because greenhouse critics focused on the fact that different climate models "predicted" a range of different results for any specific region, and this had been used to denigrate any conclusions based on climate models. But by examining the overall hydrologic processes, the GISS researchers believed they had obtained general conclusions that had dramatic implications and needed to be communicated to the policymakers and the public.

7. From a quote contained in the Combined News Service wire report of May 9, 1989.

8. *The New York Times* would later print Hansen's reply under the modified title of "Wolf in the Greenhouse," to his consternation.

CHAPTER 12

1. In medicine we have come light-years in educating the average citizen to the basics. In the more esoteric sciences—thanks in part in the United States to a "dummied-down" public education system that honors mediocrity instead of excellence—we have a largely uneducated population struggling to catch up. That race will be one of the most critical challenges faced by any society.

2. The article (which has as its subtitle "A Classic Case of Overreaction") is, in the opinion of this author, a classic example of biased journalism at its worst. With a snarling and vicious tone, the author, Warren T. Brookes, worked overtime not just to counter the opinions and findings of scientists publicly concerned about global warming, but to excoriate them and vilify their intentions, the thinly veiled allegation being that anyone who expresses concern that the United States should do something *now* about CO_2 discharges is a wild-eyed environmental freak opposed to not only the free market but, by inference, God, motherhood, apple pie, and the American way. The entire crux of the article is revealed in the following passage: "President Bush wisely told reporters: 'You can't take a policy and drive it to the extreme and say to every country around the world, "You aren't going to grow at all." That is the central issue of the global-warming debate, and it explains why the U.S. and Japanese position was supported by some 30 other developing nations which see that, just as Marxism is giving way to markets, the political "greens" seem determined to put the world economy back into the red, using the greenhouse effect to stop unfettered market-based economic expansion.'"

This sort of irresponsible name-calling does not belong in a national publication of merit. It is McCarthyesque in its intent to smear the reputation of anyone opposing free markets with the aims of Marxism and the so-called greens. It is, unfortunately, a very familiar cry of hysterical last-gasp defense when reason and logic fail. I heard much the same type of stupidity for years following publication of *Blind Trust* (William Morrow, 1986), in which I exposed the pure free-market, deregulation of the airline industry as having some curable negative pressures on the level of air safety. One extreme right wing publication promptly branded me an enemy of the free market, never taking the time to discover that this "anti-free-market advocate of big government" was also president of a bus company that had happily taken advantage of nationwide bus deregulation and secured a forty-eight state operational authority for the asking, of which I made no secret. The method of branding a reasoned demurrer from some current policy or position as being anti-American, anti-free market, or other such epithets of grossly irresponsible mudslinging does nothing to enhance reasonable debate or reasonable solutions.

INDEX

Abplanalp, Robert, 302n
Aeronomy Lab, see NOAA
Alaska, 178
Alberta, Canada, 132, 247
Albritton, Dr. Dan (of NOAA), 144, 172, 181, 243
Alliance for responsible CFC policy, 143, 183
Allied Chemical, 301n
American Academy for the Advancement of Science, 224
American Cancer Society, 316n, 317n
American Chemical Society, 48
  Meeting of September 1973, 47
  News manager (Dorothy Smith), 48
American Revolution, the, 240
American Scientist, 305n, 306n
Amundsen, Roald (Antarctic explorer), 93
Amundsen-Scott Station (South Pole, Antarctica), 168
Anderson, Dr. James (of Harvard), 59, 110–113, 136, 138, 147, 150, 151, 154, 155, 158, 160, 161, 169–171, 180, 181, 182, 284(A1), 288(A2), 309n, 310n, 311n, 313n
Antarctica, 17, 21–25, 31, 69, 70, 78, 79, 88–90, 92–106, 108, 113, 122, 130, 136, 137, 142, 145, 149, 150, 154, 155, 161, 169, 171, 173, 174, 177, 178, 182, 183, 188, 190, 195, 239, 240, 241, 252, 270, 283(A1), 299n, 303n, 305n, 310n, 311n, 313n, 317n
Antarctic Airborne Expedition of 1987, 112–114, 145, 150, 164, 165, 170, 181, 188, 196, 289(A2), 313n
Antarctic vortex, 75, 80, 100, 111, 136, 152, 154, 155, 157, 158, 161, 169, 177, 310n, 311n, 313n
Arctic Airborne Expedition of 1989, 198
Arctic vortex, 178, 311n
Argentina, 139
Arkin, Dr. P. A., 317n
Arosa, Switzerland, Dobson monitoring station, 84, 187, 313n
Arrhenius, Svante, 235, 241, 264
Aspen, CO, 183
Aurora Australis (Southern Lights), 73
Australia, 309n

Baltimore, MD, 224
Barnett, Richard, 143
Barrilleaux, Jim, 310n
Baucus, Sen. Max (Montana), 190, 192, 312n
Benedick, Richard (U.S. negotiator to Vienna Convention, 1985), 120, 144, 317n
Berkeley, University of California at, 14, 35, 42, 55, 65, 67, 116
  Chemistry department of, 37
Blake, Dr. Donald, 201
Blind Trust (book on airline safety and human factors by John Nance), 318n
Boulder, CO, 224, 284(A1), 288(A2), 307n
Braniff International Airlines, 138, 149

Branstator, Dr. Grant, 317n
British Antarctic Survey, 26–32, 69, 74, 76, 79, 80, 284(A1), 290(A2), 299n
BrO, 151, 310n
Brodeur, Paul, 305n, 306n
Broecker, Dr. Wallace S., 269
Bromley, Dr. D. Allen, 271–273
Brooks, Dr. Warren T., 318n
Bumpers, Sen. Dale (Arkansas), 192
Burford, Anne M. (administrator of the EPA), 119, 120, 140, 303n
  Book by, 64
Bush, George, 262, 263, 271–273, 318n
  Administration of, 271–273

Calcium carbonate, 235
Callis, Dr. Linwood, 109
Caltech (California Institute of Technology), 233, 236
Cambridge University (Cambridge, England), 29, 73, 197, 284(A1)
Canada, 120, 178, 194, 259, 274, 308n, 314n
Carter, James Earl "Jimmy," 119, 234
  Administration of, 117
CBC (Canadian Broadcasting System), 181
CFC-11 (chlorofluorocarbon product), 122
CFC-12 (chlorofluorocarbons), 122
CH_4 (methane), 124, 134, 201, 210, 230, 233, 234, 245
Chafee, Sen. John (Rhode Island), 123–128, 134, 140, 205, 208
Challenger disaster (1986), 186
Chemical Manufacturers Association, 101, 122, 136, 194
Chemical theory of ozone depletion over Antarctica, 98, 106, 108, 154, 169, 170, 174, 176, 182, 184, 197, 311n
Chicago, IL, 241
Chicken Little, 53, 123, 192
Chile, 139, 164, 171, 172, 183
China, 194, 308n
Chlorine nitrate, 62, 87, 302n, 305n
Christchurch, New Zealand, 90, 95, 100, 164, 165, 171
Chubachi, Dr. S., 72
Cicerone, Dr. Ralph, 46–48, 50, 51, 54, 56, 62, 63, 65, 82, 301n, 302n, 305n
Clean Air Act, 56
ClO (chlorine monoxide), 40, 59, 87, 111, 113, 138, 139, 149, 151, 154, 158, 161, 162, 165, 167, 168–170, 174, 175, 182, 299n, 300n, 303n, 305n, 307n, 310n, 311n
$ClONO_2$, 81, 303n, 305n, 306n, 307n
ClOOCl (doublet of chlorine monoxide), 23, 87, 176, 284(A1)
CNN, 181, 274
CO_2, 76, 124–126, 134, 135, 201, 210–215, 228, 230, 232–242, 245, 246, 257, 265, 270, 273, 314n, 316n, 318n
Coal News, 244
Coleman, Phil, 311n

[319]

[320] Index

Colorado, 215
Columbia University, 131, 223, 284(A1), 314n
Combined News Service, 318n
Committee on Inadvertent Modification of the Atmosphere (IMOS), 54–56, 302n
Condon, Dr. Estelle (of NASA), 137–139
Consumer Product Safety Commission (U.S.), 302n
Cornell University, 226
Crutzen, Dr. Paul, 62, 181, 301n, 302n

Danielsen, Ed, 312n
DC-8-62 of NASA (formerly of Braniff), 138, 149, 162, 168, 169, 171, 172, 309n
DelGenio, Dr. Tony, 203
Department of Commerce (U.S.), 58
Department of Energy (U.S.), 141, 212, 213, 244
Department of Health, Education and Welfare (U.S.), 302n
Department of State (U.S.), 141
De Zafra, Dr. Robert, 165, 167, 308n, 311n
Dimer, see ClOOCl
Diode array spectrometer, 100
DNA molecule, effect of UV-B on, 35, 275, 300n, 316n
Dobson instruments, 30, 69, 70, 299n, 303n
Dobson network, 82, 83, 187, 193, 194, 305n, 311n
Dobson Unit(s), 31, 83, 155, 161, 299n
Dotto, Lydia (co-author of *The Ozone Wars*), 49, 299n
"Draft protocol" (the Nordic Annex), 120
Drake Passage, 148
Du Pont Chemical Company, 37, 45, 46, 51, 52, 57, 122, 142, 183, 189, 190–192, 194–196, 300n, 301n, 302n, 308n, 311n, 312n
Durant, Will, 305n
Durenberger, Sen. Dave, 190, 192, 312n
Dust bowl of the 1930's, 215
Dynamics theory of ozone depletion over Antarctica, 98, 107, 108, 154, 165, 169, 170, 174, 176, 182, 184, 197, 307n, 311n

Einstein, Albert, 304n
El Chichón (Mexican volcano), 84, 85, 187, 199, 311n
El Niño, 259
Energy Daily, 244
Environmental Commission of the European Community, 308n
Environmental Defense Fund, 120, 317n
Environmental Pollution Subcommittee of the U.S. Senate, 123, 127
  Hearing of, June 10, 1986, see Hearing, June 10, 1986
Environmental Protection Agency (EPA) (U.S.), 58, 64, 117, 119, 120, 122, 123, 132, 139, 140, 185, 186, 194, 214, 302n
  Burford, Anne (administrator), 64
  Phase One of chlorofluorocarbon regulations, 117
  Phase Two of chlorofluorocarbon regulations, 117
ER-2 (NASA aircraft) (Lockheed ER-2/U-2), 111, 112, 136, 137, 147–149, 150, 153–161, 155, 156, 158–161, 168, 169, 171, 172, 176, 309n, 310n
Esselen Award of the American Chemical Society, 283(A1)
Estes Park, CO, 181
European Commission, the, 118
European Economic Community (EEC), 117, 120

Fahey, Dr. David, 176, 310n
Farman, Dr. Joe, 25–33, 69–72, 74, 120, 168, 181, 234, 284(A1), 290(A2), 299n, 303n, 305n, 307n, 308n, 311n
  Paper of 1984 in *Nature*, 65, 73, 76–78, 82, 84, 88, 97, 121
Farmer, Dr. Barney, 90, 100
  Barney's Barn, McMurdo Sound Antarctica, 101, 104
Federal Aviation Administration, 117, 118
Feldafing, Germany, 306n
Finland, 120, 178
Fitzpatrick, Donna, 244
Fitzwater, Marlin, 263
Flagstaff, AZ, 222
Flatirons, Boulder, CO, 78
Food and Drug Administration (FDA) (U.S.), 62, 116, 302n
Forbes magazine, 278
Ford, Gerald R., 54
Fourier, Jean Baptiste, 234
France, 314n
Freon (Du Pont trademark for principal chlorofluorocarbon product), 37, 44, 301n

Gaia, theory of, 48, 49, 77
Garcia, Dr. Rolando (of NCAR), 80, 86, 88, 306n
  Paper with Solomon, Rowland, and Wuebbles, 99
Gardiner, B. G., 69
GCM's, see General Circulation Models
General Circulation Models (GCM's), 265
Geneva, Switzerland, 76, 139, 144
German Ministry of Science, 118
Germany, 314n
Glacial/interglacial periods, 241
Global warming, 124–135, 202, 213, 215, 216, 220, 223, 227, 228, 242, 243, 249, 256, 257, 259, 260, 264, 266, 272, 273, 278, 281, 292(A3)
  Rising sea levels as a result of, 133
Goddard Institute for Space Studies, see National Aeronautics and Space Administration: GISS
Goebbels, Joseph, 275
Goldilocks effect, the, 231, 245
Goldstone, CA (NASA tracking station), 153
Gore, Cong. (later Sen.) Al (Tennessee), 131, 192, 213, 262, 309n
Graham, William (NASA administrator), 141, 194, 312n
Graves, Peter, 59
Greenhouse effect, 124–135, 205–207, 213–215, 220, 222, 223, 227–229, 234, 243,

Index [321]

245, 247, 248, 249, 251, 254, 257, 259–262, 264, 266, 267, 269, 281, 292(A3), 309n, 316n
Greenland, 178, 188, 189, 238–240
Grotch, Dr. Stanley, 316n

Halley Bay (British Antarctic Survey outpost), 25–28, 30–33, 69, 72, 74, 76, 77, 79, 80, 82, 83, 88, 121, 138, 168
Hansen, Dr. James (of NASA Goddard Institute for Space Studies), 128, 130–132, 134, 135, 203–224, 227, 228, 243, 247, 250, 251, 253, 254–262, 264, 267, 270, 272, 273, 276, 277, 279, 284(A1), 288(A2), 308n, 309n, 313n–317n
Harris, Dr. Neil, 84, 187, 194, 305n, 313n
Hartmann, Dr. Dennis, 312n
Harvard University, 48, 57, 111, 112, 138, 189, 210, 226, 284(A1), 288(A2), 301n
Hawaii, 236
HCFC's, 196, 132n, 308n
HCl, 72, 81, 87, 88, 176, 303n, 305n, 306n, 307n
Hearing, June 10–11, 1986, Senate Environmental Pollution Subcommittee, 123, 125, 127, 131, 139, 141, 215
Hearing, June 23, 1988, Senate Energy and Natural Resources Committee, 203–215, 255
Hearing, October 1987, Senate Committee on the Environment and Public Works, 189
Hearing, December 11, 1974, House Subcommittee on Public Health and the Environment, 51–53
Heath, Dr. Donald, 78–80, 144, 145, 181, 186, 193, 304n, 305n, 309n, 311n
Heckert, Richard E. (Chairman, Du Pont), 190, 195, 312n
Heterogeneous reactions, 85, 86, 122, 187, 195, 197–199, 305n, 311n, 313n
Hitler, Adolf, 274
Hofmann, Dr. David, 90, 104, 138, 168, 199, 308n
Hotel Cabo de Horno (Punta Arenas, Chile), 151, 155
Hussein, Saddam, 274

Ice core samples, 238–241, 245
Iceland, 178
Idaho, 138
Illinois Institute of Technology, 68
IMOS, *see* Committee on Inadvertent Modification of the Atmosphere
Industrial Revolution, the, 134, 245
Inn at Estes Park, 182
Intergovernmental Panel on Climate Change (UN), 271
International Geophysical Year (IGY) (1957), 27, 234, 236, 238
Ireland, 148
Irvine, University of California at, 14, 33, 36, 37, 44, 50, 60, 64, 75, 128, 146, 283(A1), 287(A2), 300n, 302n
ITT Antarctic Services Company, 94, 96, 99, 102, 163

Jakoubek, Dr. Roger, 189, 311n
Japan, 194, 308n, 314n
Japan Prize, 1989, 283(A1)
Jet Propulsion Laboratory (JPL) (Pasadena, California), 64, 75, 86, 90, 116, 308n
Jetstream, 259
Johnson, Kelly (Lockheed aircraft designer), 137
Johnson's Wax (end of production of CFC spray aerosol cans), 57
Johnston, Dr. Harold, 14, 35, 36, 42, 43, 47, 49, 55, 65, 67, 116, 222, 301n
Johnston, Sen. J. Bennett (Louisiana), 215

Kaiser Industries, Inc., 301n
Kansas, 132
Karl, Dr. Tom, 267, 277
Kaufman, Fred, 305n
Keeling, Dr. C. David (of Caltech), 234, 236, 237, 240
Keeling's Curve, 237, 240
Kellogg, Will, 315n
Kelly, Dr. Ken, 176
Kohl, Helmut (chancellor of Germany), 273
Koppel, Ted (of ABC's *Nightline*), 203
Krakatoa, 177, 199, 200
Krueger, Dr. Arlin, 304n, 309n, 311n
Kyoto, Japan, 301n

Lake Estes, 182
Laramie, University of Wyoming at, 90, 100
Lawrence Livermore Laboratory, 86, 306n, 316n
Les Diablerets, Switzerland, 76, 308n
Lindzen, Dr. Richard (of MIT), 227, 251, 257, 265, 266, 277
Lockheed Aircraft Corporation, 139, 150
Lockheed C-130 Hercules aircraft, 92, 95, 96, 100
Lockheed C-141 Starlifter aircraft, 96, 104
Lockheed Constellation, 137
Lockheed F-104 Starfighter, 137
Lockheed SR-71 Blackbird, 137
Lockheed U-2, *see* ER-2
Los Angeles, 149
Lovelock, Dr. James, 48, 49, 77, 304n, 308n

McCarthy, Raymond (Du Pont Freon division manager), 52
McCarthy, Sen. Joseph, 274
McDonald's (restaurants), 144
McElroy, Dr. Mike, 48–50, 57, 81, 181, 189, 302n
McFarland, Dr. Mack, 181, 183, 186, 187, 192, 194–196, 311n
McIntyre, Dr. Mike, 197
McMurdo Sound, Antarctica, 89, 92–97, 102–106, 108, 109, 113, 114, 136, 138, 154, 163–165, 167, 168, 171, 172, 174–176, 189
Mahlman, Dr. Jerry, 197
Major, John (prime minister of the UK), 273
Manhattan, 261, 284(A1), 288(A2)
Manhattan Project, the, 114
Margitan, Jim, 148
Mars, 314n

[322] Index

Mauna Loa (Hawaiian volcano), 236, 237
Mead, Margaret, 226, 227, 315n
Menlo Park, CA, 198
Meriwether, Dr. John (of the University of Michigan), 188
Meteorological theory of ozone destruction, *see* Dynamics theory of ozone depletion over Antarctica
Mexico, 259
Michaels, Patrick, 266, 267, 277
Military Airlift Command of the U.S. Air Force (MAC), 96
Minihole over Palmer Peninsula, 154, 158, 168
Minnesota, 247
Mississippi River, 215
MIT (Massachusetts Institute of Technology), 283(A1), 287(A2)
Mitterrand, François (president of France), 273
Moffett Field, CA, *see* National Aeronautics and Space Administration: AMES Research Center
Molina, Dr. Mario, 14, 31, 32, 34, 36–39, 41–50, 52, 54, 56, 57, 60–65, 70, 75, 77, 78, 82, 85–88, 129, 176, 177, 181, 197, 200, 222, 283(A1), 287(A2), 288(A2), 300n, 301n, 304n, 305n, 307n, 313n
Montreal, Canada, 144, 145
Montreal Agreement, September 14, 1987, 145, 146, 176, 183, 185, 189, 190, 193, 194, 317n
Mount, Dr. George, 189, 311n
Muir, Warren, 302n
Munich, Germany, 306n
Murgatroyd, Dr. Robert "Bob," 26
Murphy, Dan, 312n

N_2O (nitrous oxide), 201, 210, 233, 245, 3 0n
NASA, *see* National Aeronautics and Space Administration
National Academy of Sciences, 49, 55, 56, 58–62, 64, 117, 234, 251, 264, 304n, 308n
    Panel appointed in 1975 to determine validity of Rowland-Molina theory, 58, 60, 62
    Periodic reports on atmospheric ozone depletion, 76
    Report of NAS panel on Rowland-Molina theory validity, 58, 61, 62
National Aeronautics and Space Administration (NASA), 70, 78, 90, 110, 116–118, 128, 130, 137–139, 148, 151, 171, 173, 181, 186, 187, 197, 204, 206, 208–210, 212, 217–220, 243, 249, 250, 257, 260
    AMES Research Center (Moffett Field, CA), 136, 137, 139, 149, 150, 151, 155
    GISS (Goddard Institute for Space Studies), 88, 116, 130, 203, 204, 206, 208, 210, 212–215, 223, 224, 227, 255, 261, 262, 265, 269, 284(A1), 288(A2), 293(A3), 294(A3), 301n, 305n, 308n, 309n, 314n, 315n, 316–318n
    Goddard Space Flight Center (Greenbelt, MD), 78, 89, 153, 172, 210, 212, 305n
    Nimbus-7 satellite, *see* Nimbus-7

National Cancer Institute, 191
National Center for Atmospheric Research (Boulder, CO), 59, 67, 100, 219, 222, 224, 225, 259, 265, 284(A1), 316n
    Atmospheric chemistry division of, 78
National Climate Data Center of Asheville, NC, 267
National Energy Policy Act of 1988, 242, 253
National Environmental Research Council, 29
National Forest Act, 111
National Oceanic and Atmospheric Administration (NOAA), 59, 65, 100, 118, 138, 151, 183, 206, 288(A2), 293(A3)
    Aeronomy Lab, Boulder, CO, 59, 67, 91, 153, 164, 181, 189, 243, 284(A1), 285(A1), 287(A2), 289(A2)
    Boulder, CO, facility, 59, 65, 89
National Research Council, 118, 214
National Science Foundation, 90, 97, 100, 106, 118, 190
National Scientific Balloon Facility, 59
Natural Resources Defense Council (NRDC), 120, 185
Nature (British scientific journal), 43, 46, 69, 74, 78, 293(A3), 302n, 307n
Nebraska, 132
Newcomb-Cleveland prize of the American Association for the Advancement of Science, 283(A1)
New Scientist, The, 304n
New York City, 223, 241, 254, 284(A1)
New Yorker, The, 305n, 306n
New York Times, The, 48, 107, 213, 218, 224, 225, 254, 258, 267, 316n, 318n
New Zealand, 22, 85
Nightline (ABC news), 203
Nimbus-7 (satellite), 24, 78, 145, 152, 153, 155, 158, 161, 186, 187, 193, 304n
    SBUV instrument aboard, 79
    TOMS (Total Ozone Mapping Spectrometer) aboard Nimbus-7, 79, 135, 152, 153, 168, 177, 178, 309n
Nitric acid, 157
Nixon, Richard M., 35, 302n
NO (nitric oxide), 300n, 303n
NO_2, 76, 85, 86, 88, 104, 124, 300n, 303n, 305n, 307n
NOAA, *see* National Oceanic and Atmospheric Administration
Nordic Annex, *see* "Draft protocol"
Northwestern University, 236
Northwest passage, the, 213
Norway, 120, 178
NOZE-1 (National Ozone Expedition #1 [1987]), 94, 96, 99, 100, 113–115, 118, 121, 122, 138, 144, 164, 165, 173, 307n, 308n, 311n, 313n
NOZE-2 (National Ozone Expedition #2), 119, 145, 146, 168, 171, 172, 175, 176, 311n, 313n

OClO (chlorine dioxide), 89, 99, 103, 104, 164–167, 174, 189, 307n, 311n
Office of Management and Budget (OMB) (of the White House), 131, 140, 207–209, 260–264, 309n

Index [323]

Office of Science and Technology Policy (of the White House), 271
Oregon, 58
Ormund, Flo, 289(A2)
Oslo, Sweden, 119
Ozone Trends Panel, 145, 186–188, 193–196, 293(A3), 311n, 312n, 316n
Ozone wars, 14, 17, 18, 28, 32

Paleoclimatology, 239, 240
Palestine, TX, 112
Palmer Peninsula, 150, 161, 164, 168
Palmer Station, Antarctica, 155, 168
Panama, 155
Parfait, Michael (author of *Southlight*), 96
Paurowski, Carolyn, 255
Pend Oreille Lake, ID, 110–111
Pennwalt Corporation, 301n
Peterson, Russell, 127
Phase One and Phase Two of EPA regulations on chlorofluorocarbons, 117, 185
Phaseout of nonessential CFC aerosol products by U.S government, 63
Photodissociation, 39
Pimentel, Dr. George, 300n
Pioneer (spacecraft), 210, 314n
Polar Ozone Workshop, May 9–13, 1988 (Snowmass, CO), *see* Snowmass Village, CO
Polar Stratospheric Clouds (PSC's), 73, 86, 87, 152, 154, 157–158, 170, 171, 176–178, 198, 199, 284(A1), 305n, 311n, 313n
Polar vortex, *see* Antarctic vortex; Arctic vortex
Precision Valve Corporation, 302n
Press, Dr. Frank, 234
Prince Albert Mountains (of the Scott Coast of Antarctica), 93
Proffitt, Dr. Mike, 312n
PSC's, *see* Polar Stratospheric Clouds
Puerto Montt, Chile, 155
Punta Arenas, Chile, 139, 146–148, 151–153, 155, 158, 161, 164, 165, 167, 168, 171–173, 176, 181–185, 188, 189, 289(A2), 310n, 313n

Queen Mary College (of London University), 115

Racon, 301n
Ramanthan, Dr. V., 210
Rasool, Dr. Ichtiague, 223
Reagan, Ronald W., 64, 121, 142, 242, 301n
  Administration of, 82, 119, 140, 143, 185, 208, 212–214
  Policies of, 140, 141
Revelle, Dr. Roger, 235, 316n
"Revelle effect," the, 236
Rind, Dr. David, 203
Roan, Sharon (author of *Ozone Crisis*), 292(A2), 301n, 308n, 312n
Roosevelt, Theodore "Teddy," 111
Ross Ice Shelf (Antarctica), 93
Rowland, Dr. Sherwood "Sherry" F., 14, 31–34, 36, 37, 39, 41–50, 52, 54, 56–58, 60–65, 70, 75–78, 81–86, 88, 115, 121, 128–130, 134, 135, 143–146, 177–181, 186–189, 192, 194, 199, 200–202, 207, 215, 222, 228, 270, 276, 277, 279, 283(A1), 287(A2), 288(A2), 300n–302n, 304n, 305n, 308n, 313n
Paper with Garcia and Wuebbles, 99
Rowland, Joan, 43
Rowland-Molina basic theory of catalytic chlorine destruction of ozone, 40, 44, 45, 47, 49–62, 64, 65, 70, 76, 77, 81, 83, 84, 113, 115, 299n, 300n, 302n–304n, 313n
Rowland-Molina paper (1974), 32, 43, 47–49, 74, 115, 129, 300n–302n
Ruckelshaus, William (former secretary of the interior), 119
Russia, *see* U.S.S.R.

Sagan, Carl, 226, 228, 276, 277
Salter, Peter, 152
Sanders, Dr. Ryan, 189, 311n
Santiago, Chile, 149
Sato, Dr. Haruo, 306n
SBUV, *see* Nimbus-7
Schiff, Harold (co-author of *The Ozone Wars*), 49, 293(A3), 305n
Schmeltekopf, Dr. Art, 89, 91, 102, 103, 138, 164–166, 311n
Schneider, Dr. Stephen (of NCAR), 218, 222–229, 242–244, 246, 247, 250, 251, 253, 265, 266, 270, 276–279, 284(A1), 289(A2), 293(A3), 315n
Schoeberl, Dr. Mark, 108, 164, 181, 197, 309n, 312n
Science magazine, 47, 48, 191, 213, 283(A1), 293(A3), 302n, 307n, 309n, 311n, 316n, 317n
Scientific American, 277
Scotland, 148
Scott, Robert F., 93
Seattle, WA, 241, 288(A2)
Senate Committee on Commerce, Science and Transportation (Subcommittee on Science, Technology, and Space), 260
Shabecoff, Phil, 218
Shanklin, J. D., 69
Shapiro, Irving, 191
Sierra Club, 50
Skin cancer, resulting from UV-B radiation exposure, 35, 48, 178, 303n, 304n, 316n
Skunk Works (Lockheed Aircraft design unit), 137
Smith, Dorothy (news manager, American Chemical Society), 48
Snowmass Village, CO, 179–183, 186, 196, 198–202, 287(A2), 289(A2), 313n
Solar theory of ozone depletion over Antarctica, 98, 103, 109
Solomon, Dr. Phil, 165
Solomon, Dr. Susan, 20, 36, 65–68, 72–74, 76–78, 80, 81, 84–86, 88, 91–94, 96–106, 108, 109, 113, 114, 138, 143, 154, 163–167, 171, 172, 176, 178, 181, 188, 197–199, 222, 276, 284(A1), 287(A2), 288(A2), 305n, 306n, 307n, 308n, 311n

[324] *Index*

Susan Solomon Rabbit's Foot Effect, 104, 165,
Sorbonne (France), 300n
Southlight (book by Michael Parfait), 96
Space shuttle, NASA, 42, 301n
Spectrophotometer, 30, 299n
Spencer, Dr. John, 60, 302n, 305n
SRI International (Menlo Park, CA), 198, 307n
SST (supersonic transport aircraft project), 34, 35, 42, 55, 116, 301n
Stafford, Sen. Robert, 189, 190, 192, 312n
Stalin, Joseph, 274
Stapleton Airport (Denver), 242
State University of New York at Stony Brook, 91, 104
Stolarski, Dr. Richard, 46–48, 50, 82, 181, 197, 301n, 309n
Stony Book, *see* State University of New York at Stony Brook
Straits of Magellan, 139, 148
Stratospheric Ozone Protection Plan of the EPA, *see* Phase One and Phase Two of EPA regulations on chlorofluorocarbons
Strobach, Donald, 308n
Suess, Dr. Hans, 235
Sulfuric acid aerosols, 199
Sullivan, Walter (of *The New York Times*), 48, 107, 218, 224, 225
Sweden, 120, 178, 301n
Switzerland, 300n, 303n
  Swiss Alps, 240
Syowa, Antarctica (Japanese research station), 72

Teller, Dr. Clifford, 222
Thatcher, Margaret (former prime minister of the UK), 273
Theory of Plate Tectonics, 234
Thomas, Lee (EPA administrator), 119, 120, 140, 141, 144, 185
Thule Air Force Base (Greenland), 188, 189, 198, 311n
Tierra del Fuego, 148
Titanic, the, 82
Tolbert, Dr. Margaret A. "Maggie," 198, 199, 313n
TOMS (Total Ozone Mapping Spectrometer), *see* Nimbus-7
Tom's Restaurant (Manhattan, NY), 130, 131, 284(A1)
Toon, Dr. Geoff, 176
Toronto, Canada, 223, 227, 228
Trenberth, Dr. Kevin, 317n
Tso, Dr. T. L., 307n
Tuck, Dr. Adrian (of NOAA), 151, 153, 160, 168, 181–183, 197, 285(A1), 289(A2), 311n, 312n
Tung, Dr. Ka Kit, 75, 76, 197
Turbulent diffusion, 38
Twain, Mark, 224
Tyndall, John, 234

United Kingdom, 148, 273, 274, 314n

United Nations, 271
United Nations Environmental Programme, 118, 119, 142, 308n
University of California at Irvine, *see* Irvine, University of California at
University of California at Riverside, 306n
University of Freiburg (Germany), 300n
University of Iowa, 210, 313n
University of Kyoto (Japan), 210
University of Maryland, 116
University of Mexico, 300n
University of Michigan, 46, 47, 50, 59, 188, 301n
University of Wyoming at Laramie, *see* Laramie, University of Wyoming at
USA Today, 280
U.S.S.R., 193, 194
UV-B (ultraviolet-B) radiation, 22, 23, 28, 39, 46, 95, 177, 178, 191, 194, 232, 258, 299n, 300n, 302n, 303n, 311n, 312n, 316n, 317n
  Effect on DNA molecule, 25
  Link to skin cancer, 35

Van Allen, Prof. James, 313n
Venus, 124, 130, 313n, 314n
Vienna, Austria, 41, 42, 44, 301n, 302n, 308n
Vienna Accord, *see* Vienna Convention on CFC's and ozone
Vienna Convention on CFC's and ozone (March 1985), 120, 121, 144
Vostok (Soviet Antarctic research station), 240

Wade, Nicholas (of *The New York Times*), 255, 256, 258, 260, 263, 264, 316n, 317n
Wang, Dr. F. C., 307n
Washington National Airport, 242
Washington Post, The, 260, 266
Watergate scandal, 35
Watson, Dr. Robert, 88–91, 98, 110–118, 121, 123, 128, 129, 134–139, 144, 145, 148, 164, 167, 170–173, 181, 186–189, 193, 194, 207, 215, 219, 243, 248, 250–252, 269, 270, 279, 284(A1), 289(A2), 303n, 305n, 308n, 317n
Watt, James (former secretary of the interior), 119
Watt, Sir James, 240
Weiner, Jonathan (author of *The Next One Hundred Years*), 293(A3), 316n
White House, the, 54, 214, 242, 253
Wilkniss, Dr. Peter, 113, 190, 308n
Williams, Ron, 150, 155–159
Winter-fly missions to Antarctica, 90
Wirth, Sen. Tim (Colorado), 204–206, 215, 242, 244, 246, 249, 250, 253
Wofsy, Dr. Steven, 48, 50, 57, 181, 302n
World Meteorological Organization, 117, 223, 293(A3), 308n
Wuebbles, Dr. Donald, 81, 86, 88, 306n
  Paper with Solomon, Rowland, and Garcia, 99

Yung, Dr. Yuk Ling, 210